# EXERCISE PHYSIOLOGY

## AND ANATOMY OF MOVEMENT

# COLIN CLEGG

**REVISED BY**

**STEPHEN INGHAM**

Steve Ingham, through five Olympic cycles, has provided scientific support to over 1000 athletes including 50 Olympic or World medalists; amongst them some of Britain's greatest athletes: Sir Steve Redgrave, Sir Matthew Pinsent and Jessica Ennis-Hill. A graduate in Sports Science from the University of Brighton, Steve gained his PhD from the University of Surrey. He began his career as a physiologist for the English Sports Council in 1996 and later joined the British Olympic Medical Centre in 1998, working in athletics, rowing and bobsleigh. In 2005 Steve moved to the English Institute of Sport, based at Loughborough University, furthering his work in athletics. In 2009 he took on the role of Head of Physiology at the English Institute of Sport and as Director of Science and Technical Development in 2013. Steve is a Fellow of the British Association of Sports and Exercise Sciences."

Copyright © C.A. Clegg 2016
First Published by Feltham Press 1995
Reprinted by Feltham Press 1995
Reprinted by Feltham Press with additions in colour 1996
Reprinted by Feltham Press Autumn 1996
Reprinted by Feltham Press with additions 1998
Reprinted by Feltham Press with revisions 2000
Reprinted by Feltham Press 2001
Reprinted by Feltham Press 2004
Revised 2016

ISBN 978-1-904728-94-8

British Library Catalogue-in-Publication Data.

A catalogue record for this book is available
from the British Library.

# EXERCISE PHYSIOLOGY

# CONTENTS

# ACKNOWLEDGMENTS

Thanks are especially due to Dr Jo Doust of the Chelsea School of Physical Education, Sport Science, Dance and Leisure, University of Brighton, who most generously gave of his valuable time to read the entire manuscript, and who made many constructive suggestions which were of the greatest help to the author.

I would also like to thank Dr David Parry of King's College Hospital and Dr Barry Mitchell of the Anglo-European College of Chiropractic, Bournemouth, for help with the anatomy section. I would also like to thank Chris Tredwell for his work in the preparation of the graphs, Steve Mayer, and James Leaver for their help in the final stages of the preparation of the manuscript, and Chris Hanks for his encouraging comments.

I am especially grateful to Sam Denley of Feltham Press who devoted much care and attention to the creation of the artwork and to the design and production of the book as a whole, also to Steve Lucas for his assistance with the preparation of the colour plates in the anatomy section.

Finally, I would like to express my gratitude to my wife Ann, without whose help and support the entire project would not have been possible.

## ACKNOWLEDGMENTS for the revised edition 2000

The revisions by Stephen Ingham, Senior Exercise Physiologist at the British Olympic Medical Centre, bring a more applied approach to certain sections, and where necessary have brought the terminology up to date, in accordance with widely accepted practice.

## ACKNOWLEDGMENTS for the revised edition 2016

I would like to thank those who provided constructive comments on the content, Stephen Baker of eBooks by Design for his patience and professionalism preparing the book for ebook platforms; and my wife Ann for her continuing help and support.

# PREFACE

The aim of this book is to provide a concise but comprehensive introduction to the main ways in which the human body responds to the demands of exercise.

Exercise typically involves movement; but there are elements within exercise which may be static, involving tension and balance. Exercise may also be either concentrated into a short period of intense activity, or sustained over a longer time. Whatever the type of exercise, the body systems adjust and adapt to it, both in the short term and the long term – some more dramatically than others. The study of how the body adapts to the demands of exercise is known as **Exercise Physiology**.

Although the huge number of sports and activities available involve a wide variety of actions, they all have certain major interacting systems in common, and it is on these that attention will be focussed in this book.

Exercise always involves muscular activity, so that **Muscles and Bones in Action** is a good starting point for the study of exercise physiology.

The **Energy Relations** of muscular activity are critical, as without a continuous supply of energy, muscular contraction is not possible.

The increased use of energy by muscles during exercise results in an increase in the demand for fuel and oxygen, and in the production of waste products. Therefore the **Circulation**; and **Breathing, Gas Exchange and Transport** systems adapt accordingly.

Other body systems are also involved in the response to exercise, but they either do not show such specific adaptations as those mentioned above, or they may be so closely involved with them, that they are difficult to deal with independently. Therefore these systems are either dealt with as their involvement arises, or in the Chapter on **Exercise, Fitness and Health**, where consideration is also given to the changing response of the body with age.

**Training and performance applications** are highlighted in sections throughout wherever relevant; and the broader principles of training are dealt with in the Chapter on **Training Principles**.

**Key Points** of special importance are summarised at the end of each chapter.

**Photocopiable packs** for staff/student use on **Measurement and Testing of Physical Performance**, and of **Assignments and Solutions** designed to help students work through this book, encourage self-appraisal, and develop examination technique, are available to accompany this volume.

# 1 MUSCLES AND BONES IN ACTION

## OBJECTIVES

To enable the reader to understand the following.

❖ The structure of skeletal muscle.

❖ How the sliding filament mechanism works in muscle contraction.

❖ The different types of muscle fibre, and their use during exercise.

❖ Muscle fibre types as predictors of performance.

❖ The effect of training on muscle fibres.

❖ Motor units.

❖ How muscle fibres are stimulated by nerves.

❖ Muscle fatigue.

❖ The role of proprioceptors in muscle contraction.

❖ Muscles and bones in action.

## 1.1 STRUCTURE OF A SKELETAL MUSCLE

A skeletal muscle consists of bundles of muscle fibres that have a striped or striated appearance under the microscope.

The muscle fibres can be up to several centimetres long and can branch. They do not usually run from one end of the muscle to the other, and where they overlap they are bound together by connective tissue. The connective tissue sheaths that bind the muscle fibres together are continuous with each other, and ultimately, with the tendons at each end of the muscle, so that contraction of the fibres is transmitted efficiently via the tendons to the bones of the skeleton.

**Figure 1.1:** *Structure of a skeletal muscle.*

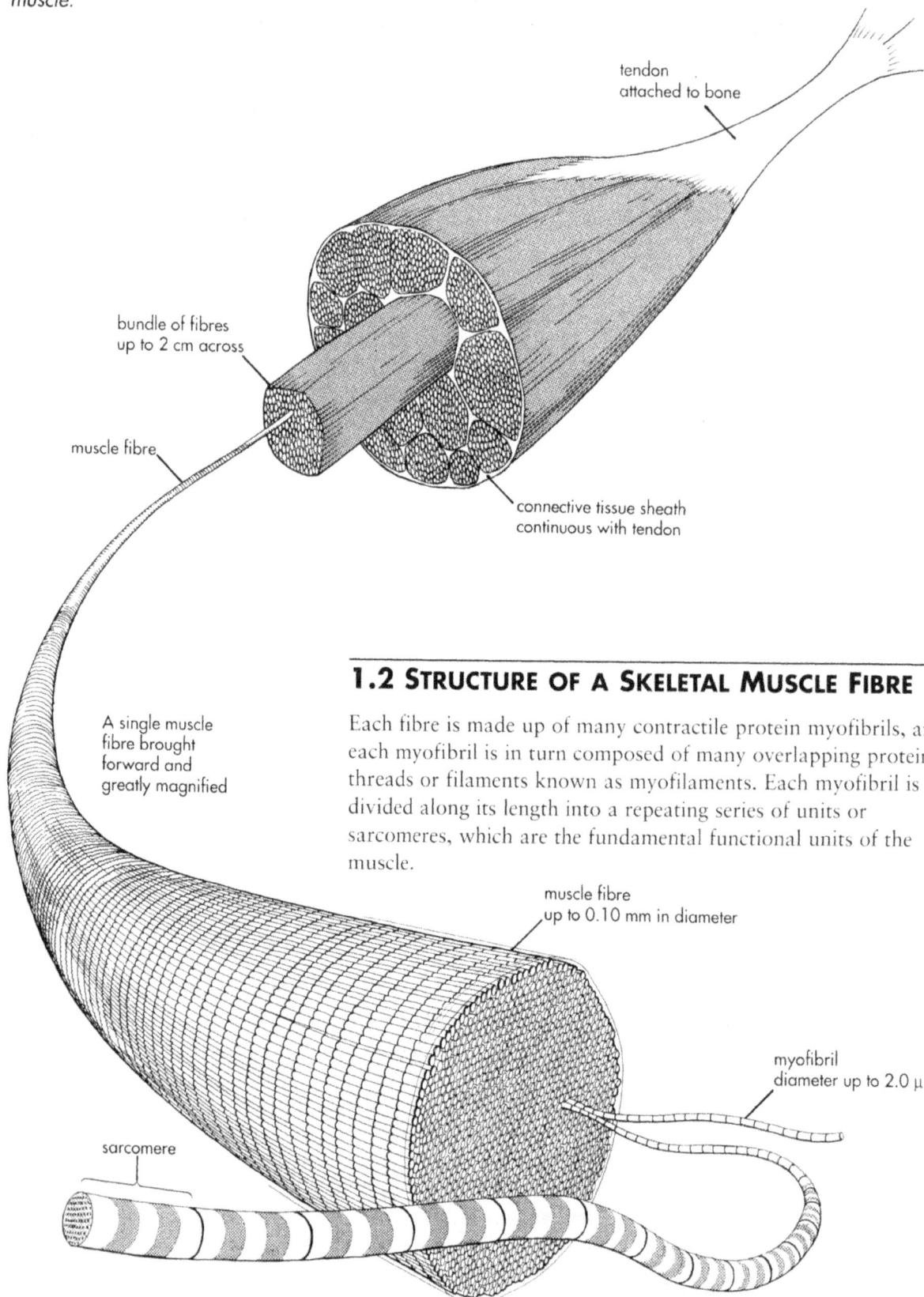

tendon
attached to bone

bundle of fibres
up to 2 cm across

muscle fibre

connective tissue sheath
continuous with tendon

A single muscle
fibre brought
forward and
greatly magnified

## 1.2 STRUCTURE OF A SKELETAL MUSCLE FIBRE

Each fibre is made up of many contractile protein myofibrils, and each myofibril is in turn composed of many overlapping protein threads or filaments known as myofilaments. Each myofibril is divided along its length into a repeating series of units or sarcomeres, which are the fundamental functional units of the muscle.

muscle fibre
up to 0.10 mm in diameter

myofibril
diameter up to 2.0 μ

sarcomere

**Figure 1.2:** *Structure of a single muscle fibre.*

## 1.3 STRUCTURE AND FUNCTION OF A SARCOMERE

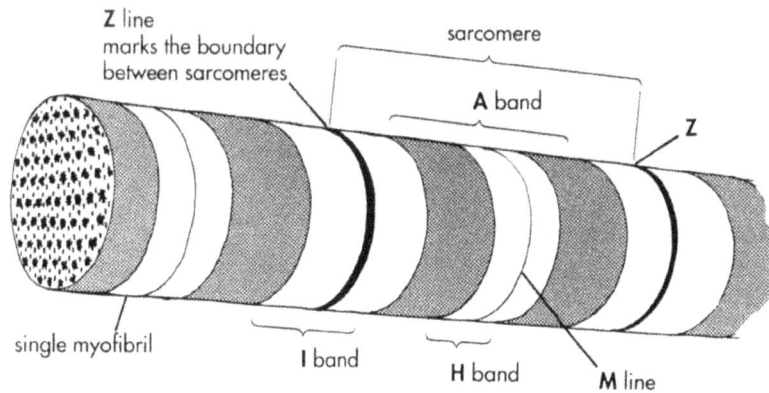

**Z** line
marks the boundary
between sarcomeres

sarcomere

**A** band

**Z**

single myofibril

**I** band

**H** band

**M** line

**Figure 1.3:** *The striated appearance of a sarcomere.*

The protein myofilaments of the sarcomere are of two main types; thin filaments of actin, and thicker filaments of myosin. The overlapping arrangement of these filaments gives rise to the striped or striated appearance of the fibres when observed under the microscope.

The sarcomere contracts as a result of these overlapping filaments sliding over each other, in what is known as the sliding filament mechanism. The simultaneous contraction of each sarcomere results in the contraction of the muscle fibre as a whole.

### 1.3.1 SLIDING FILAMENT MECHANISM

When stimulated by a motor nerve impulse, the actin filaments which are attached to the Z discs at the end of each sarcomere, are slid between the myosin filaments, drawing the Z discs closer together. This causes each sarcomere to shorten. The sliding action is operated by a series of cross bridges between the filaments. These cross bridges move the actin filaments actively one way, but not the other way on their recovery stroke, thus explaining why the sarcomere, and therefore the muscle as a whole, is only effective in contraction, and cannot actively restretch itself. Typically a contracted muscle is restretched by the contraction of an opposing muscle *(Section 1.10)*. If a muscle is overstretched, then the elasticity of its component tissues restore it to its original length. If an immovable resistance prevents the contraction of the muscle as a whole, the sarcomeres may still contract a certain amount by stretching the elastic component of the muscle instead of moving the bone(s) to which they are attached.

**Figure 1.4:** *Sliding filament mechanism.*

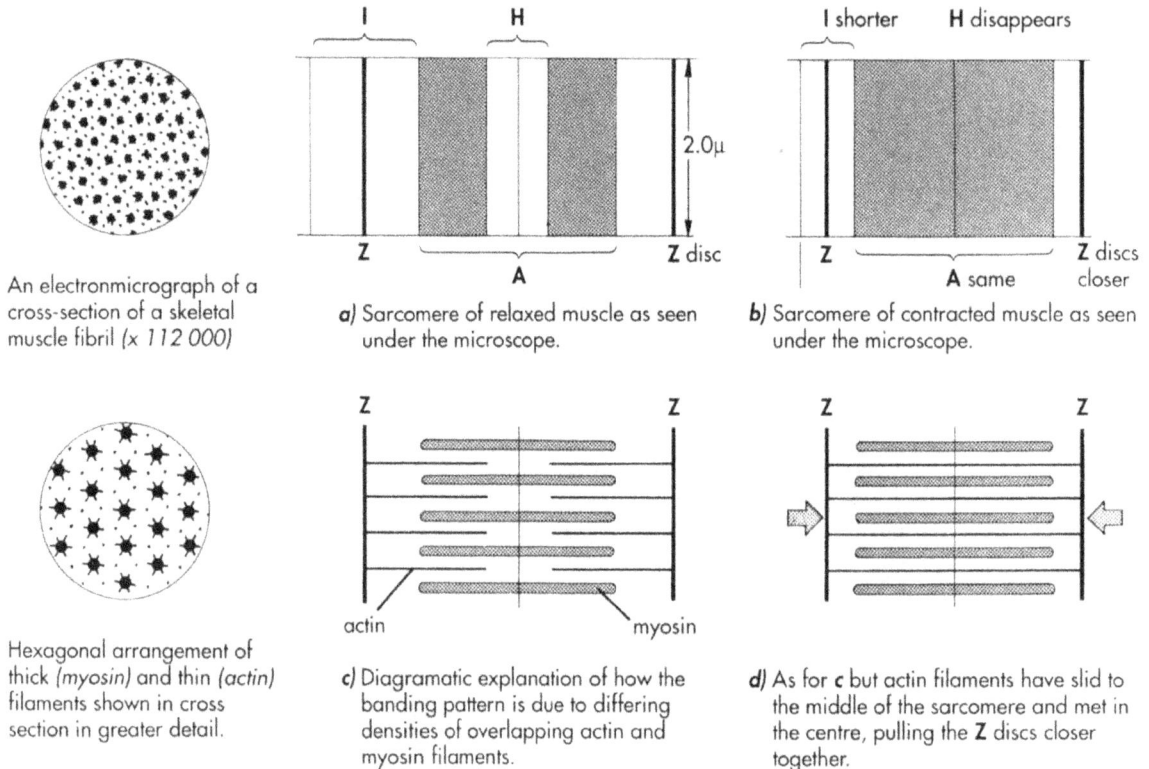

An electronmicrograph of a cross-section of a skeletal muscle fibril (x 112 000)

*a)* Sarcomere of relaxed muscle as seen under the microscope.

*b)* Sarcomere of contracted muscle as seen under the microscope.

Hexagonal arrangement of thick *(myosin)* and thin *(actin)* filaments shown in cross section in greater detail.

*c)* Diagramatic explanation of how the banding pattern is due to differing densities of overlapping actin and myosin filaments.

*d)* As for **c** but actin filaments have slid to the middle of the sarcomere and met in the centre, pulling the **Z** discs closer together.

| Sarcomere lengths | | |
|---|---|---|
| Fully extended | 3.6μ | 100% |
| At rest | 2.2μ | 59% |
| Fully contracted | 1.6μ | 44% |

The force generated by a muscle is related to the degree of overlap of the actin and myosin filaments, which in turn determines the number of cross bridges that can be formed at any one time. For example, the more a muscle is stretched past an optimum length, the smaller the force that can be exerted, as there is less and less overlap between the filaments, and fewer and fewer cross bridges that can be formed.

### 1.3.2 MOLECULAR BASIS OF SLIDING FILAMENT MECHANISM

The myosin filaments have protein projections or cross bridges or 'heads' which extend towards the actin filaments. When the muscle is at rest the myosin cross bridges are not connected to the actin filaments, and an ATP molecule is bound to the free end of each cross bridge.

The actin filaments are associated with two other proteins which are involved in control of the contraction mechanism, namely **tropomyosin** and **troponin**.

In the absence of free calcium ions, tropomyosin and troponin prevent the connection of the cross bridges between the myosin and actin filaments.

The spread of an excitatory impulse throughout the muscle fibres, causes the release of calcium ions from the sarcoplasmic reticulum *(a system of membranous sacs)* around the myofibrils. Troponin has a high affinity for calcium ions, and in their presence the inhibitory effect of tropomyosin and troponin is removed, and the myosin cross bridges connect with the actin.

**Figure 1.5:** *Diagrammatic model of cross bridge action.*

*a)* Filaments and cross bridges.

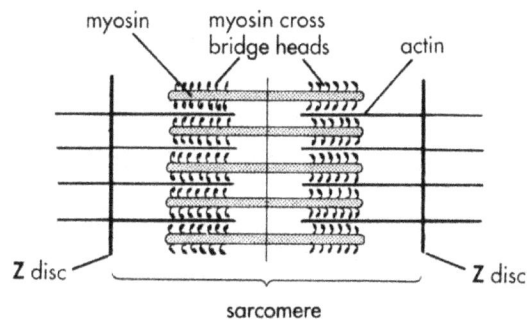

*b)* Action of cross bridges.

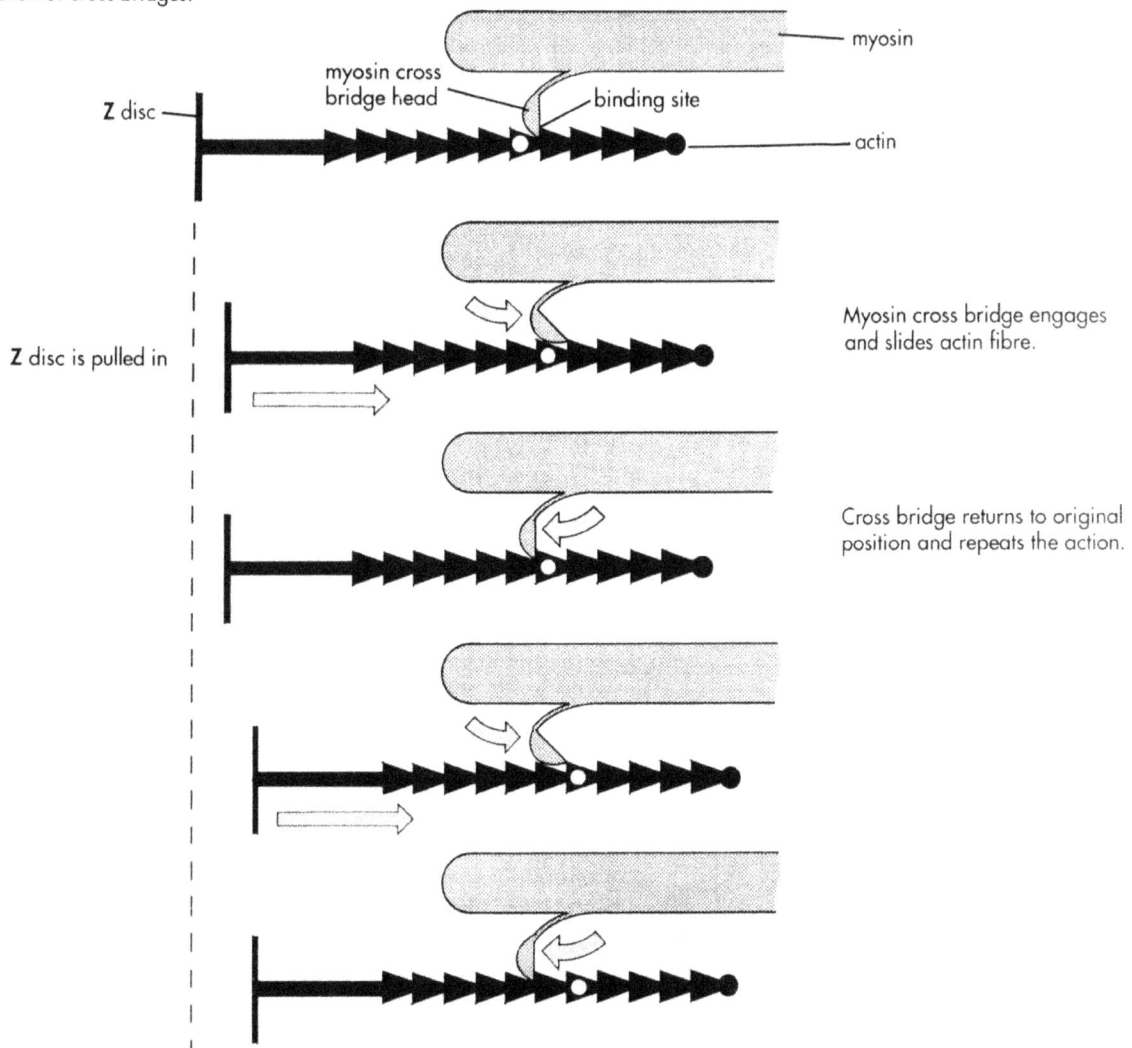

Myosin cross bridge engages and slides actin fibre.

Cross bridge returns to original position and repeats the action.

### 1.3.3 ROLE OF ATP

The coupling of actin and myosin forms an **actomyosin** complex. The formation of this complex activates an enzyme, myosin-ATPase, which catalyses the breakdown of ATP with the release of energy.

$$ATP \rightarrow ADP + P + Energy$$

The released energy is used to move the cross bridges actively in one direction, which in a normal contraction *(in which the muscle shortens)* causes the actin filaments to be slid over the myosin filaments towards the centre of the sarcomere, thus pulling the Z discs at each end of the sarcomere closer together so that the muscle contracts.

The cross bridge connection is broken by the combination of another ATP molecule with the actomyosin.

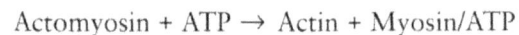

$$Actomyosin + ATP \rightarrow Actin + Myosin/ATP$$

As a result a single cross bridge may undergo many reconnections per second with the actin filament during a muscle contraction.

The cross bridges operate out of phase with each other, so that at any one time about half are attached, whilst the others are recovering, thus providing a smooth steady action.

The coupling and uncoupling of the cross bridges between the myosin and actin filaments continues as long as there are sufficient calcium ions available to inhibit the troponin-tropomyosin system (this will be the case as long as the muscle is being stimulated), and as long as ATP is available - in the absence of ATP the cross bridges remain coupled and the muscle remains in contraction (rigor).

### 1.3.4 MUSCLE RELAXATION

When the motor nerve impulse stops, the calcium ions are actively pumped back into the sarcoplasmic reticulum, and the troponin-tropomyosin system resumes the blocking of the cross bridge connections. At the same time the myosin-ATPase activity decreases and therefore ATP is not broken down and energy is not released, no tension is generated, and the muscle is relaxed.

## 1.4 TYPES OF MUSCLE FIBRE

Two main types of striated skeletal muscle can be distinguished on the basis of their speed of contraction, namely **Slow Twitch** **(ST)** and **Fast Twitch** **(FT)** fibres. Human skeletal muscles consist of mixtures of each type of fibre.

The relative proportion of each fibre type varies in the same muscles of different people, and in different muscles within the same person. For example; elite marathon runners can have a greater proportion of ST fibres in their leg muscles, and elite sprinters more FT fibres, but this is not always the case *(Section 1.4.5)*. Within the same person the soleus muscle in the calf has a greater proportion of ST fibres, possibly as it is primarily a postural muscle involved in maintaining balance over long periods of time; and the triceps muscle in the arm has a greater proportion of FT fibres.

### 1.4.1 SLOW TWITCH MUSCLE FIBRES *(ST)* or TYPE I

These have a slower contraction time than the fast twitch fibres, and they are better adapted to low intensity, long duration endurance work. They have a high capacity for aerobic respiration as a result of several modifications. For example, they have a better blood supply as a result of a greater number of capillaries per fibre than the fast twitch fibres. Slow twitch fibres have more myoglobin, which is a red muscle pigment that combines with oxygen, and for this reason they can also be called 'red fibres'. Myoglobin has a greater affinity for oxygen than haemoglobin, it thus 'robs' oxygen from the oxyhaemoglobin in the blood, and in turn acts as a store of oxygen within the muscle, releasing it to the mitochondria *(microscopic structures in the fibres which are the centres of aerobic respiration.)*. Slow twitch fibres also have more mitochondria, with their associated enzymes for aerobic respiration, than the fast twitch fibres.

These adaptations result in a greater ability to utilise oxygen, known as the **oxidative potential**, so they are also sometimes called Slow Oxidative *(SO)* fibres.

### 1.4.2 FAST TWITCH MUSCLE FIBRES *(FT)* or TYPE II

These have a fast contraction time, and are well adapted to high intensity short duration work. Such fibres have a larger diameter *(cross section)* than ST fibres due to an increased number and thickness of the myosin filaments, giving a superior mechanical arrangement from which a greater force of contraction can be generated. FT fibres are further adapted to rapid and powerful contractions through a greater rate of calcium ion *($Ca^{++}$)* control for rapid fibre activation and relaxation. These fibres possess higher concentrations of the enzyme Myosin ATPase permitting higher rates of energy release from ATP for myosin cross bridge action. FT fibres also have large stores of phosphocreatine *(section 2.4)*, increasing the potential for short-term regeneration of ATP. They have a low oxidative potential *(aerobic capacity)* having less capillaries per fibre, less myoglobin *(therefore also being known as 'white fibres')* and fewer and smaller mitochondria than ST fibres. As a result FT fibres produce ATP

**Table 1.1:** *Types of muscle fibres.*

| I | Slow Twitch *(ST)*, or Slow Oxidative *(SO)* |
|---|---|
| IIa | Fast Twitch *(FTa)* or Fast Oxidative–Glycolytic *(FOG)* |
| IIb | Fast Twitch *(FTb)* or Fast Glycolytic *(FG)* |

mainly via rapid, but inefficient, anaerobic pathways. Since the lactic acid system of anaerobic glycolysis is one of the major pathways, these fibres possess large stores of muscle glycogen and glycolytic enzymes and are therefore also known as Anaerobic or Fast Glycolytic *(FG)* fibres.

### 1.4.3 SUMMARY OF MUSCLE FIBRE TYPES

Although it is convenient to deal with muscle fibres as being either slow twitch or fast twitch, there is in fact a range of fibre types including those that have a mixture of oxidative and glycolytic properties *(Type IIa or FTa or Fast Oxidative-Glycolytic (FOG) )*, see Table 1.1.

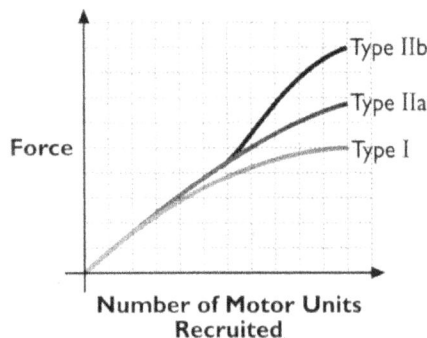

**Number of Motor Units Recruited**

❖ *Training and Performance Applications*

*In order to train the whole muscle for strength (the maximum force that can be exerted in a single effort), and power (the amount of work that can be achieved in a certain time), it must be subjected to maximum stress, thus engaging as many fibres at the same time as possible. In weight training for example, many repetitions of an easily managed load will improve muscle endurance, but only engage relatively few fibres at any one time. Whereas a few repetitions of near maximum load, will improve muscle strength and power, by engaging virtually all the muscle fibres at once. As with all training, gains are only made as a result of progressive overload, along with adequate recovery periods. Adaptations of the nervous system in increasing activation and coordination of motor units, contribute to strength gains in the early stages of training.*

❖ *Training and Performance Applications*

*The proportion of different muscle fibre types in the different muscles, which is largely inherited, explains the high specificity of training, and the specialisation of different individuals to individual sports. However, in addition to the fibres described previously, another group of relatively unspecialised fibres can be identified if special staining techniques are used. These are the Type IIc or undifferentiated fibres, which make up less than 5% of the total fibres. It is suggested that endurance training can result in Type IIb being converted into Type IIa, and that these Type IIc could be fibres in transition between the two. This would go some way in explaining why too much long steady work in running can result in a loss of speed.*

*With age the total number of fibres is reduced, with a reduction of maximum oxygen consumption and strength, which results in a dramatic drop in athletic ability.*

### 1.4.4 USE OF MUSCLE FIBRE TYPES DURING EXERCISE

All the different types of fibre are used *(recruited)* in all types of exercise. For example, although the slow twitch fibres are particularly adapted to low intensity aerobic endurance work, they are always recruited at the start of exercise, independent of the intensity of the exercise. Similarly, although the fast twitch fibres are particularly adapted to high intensity anaerobic activities involving all out rapid powerful movements, they are also progressively recruited during low intensity endurance work as fatigue increases.

However, considering a muscle where there is a gradual increase in the force of contraction, there is an overall increase in involvement of fibres from Type I *(SO)*, to IIa *(FOG)* to IIb *(FG)*.

**Table 1.2:** *Comparison of percentage fibre compositions of different categories of athlete.*

| | Range of % ST fibres | Average % ST fibres |
|---|---|---|
| Male marathon runners | 50 – 95 | 85 |
| Male 800m runners | 50 – 80 | 55 |
| Male sprinters | 20 – 55 | 35 |

## 1.4.5 MUSCLE FIBRE TYPES AS PREDICTORS OF PERFORMANCE

Broadly, the relative fibre composition of an athlete's muscle will act as a good indicator of the activities at which they are likely to excel. Research suggests that there is a certain proportion of fibre types required to perform at the top level, e.g. long distance runners between 70-90% ST fibres. Sprinters possess only between 45-60% ST fibres, thus having a greater proportion of FT fibres. These distinctions between muscle fibre composition are only clear at elite level. Fibre composition is clearly not the sole determinant of successful performance as there is an overlap of fibre composition and success within a variety of sports. The characteristics of muscle fibres e.g. mitochondrial density, and myosin ATPase activity, play an important role, above simple fibre proportions. Furthermore, other systems, including the neurological, biochemical, and hormonal, all combine to produce a physiological performance.

❖ *Training and Performance Applications*

*In man the number of fibres within a particular muscle appears to be fixed early in life. Any increase in muscle size (cross sectional area) as a result of training is due to an increase in the size of the fibres (hypertrophy) already present, rather than to an increase in the number of fibres by longitudinal division. This increase in size is due to an increase in the number and size of myofibrils per fibre, with an associated increase in the amount of proteins, especially myosin. The greater the number of fibres (which cannot be increased by training), and the greater the cross sectional area (which can be increased by training) the greater the strength of the muscle, although some increases in strength and economy are due to the central nervous system developing the capacity to recruit motor units into simultaneous contraction.*

*Training affects all fibre types in the same way. For example endurance training results in an increase in the capillary density, and in the number and size of mitochondria in all fibre types. In fact one of the most significant adaptations to endurance training is an increase in the activity of the enzymes associated with aerobic respiration in the mitochondria. The fact that successful performance is possible without the optimum proportion of fibre types, is illustrated by the fact that endurance training can raise the oxidative capacity of FT fibres even above that of ST fibres in an untrained subject. Similarly, power training results in an increase in the anaerobic potential of both fibre types. However, the cross sectional area of FT fibres increases more than that of the ST fibres with strength training. Although the aerobic and anaerobic capacities of all fibre types can be improved by training, it will be the fibre type most involved with the type of exercise that will experience the major effects of selective fatigue and training. The contraction times of the respective types remain unaltered. All training gains are rapidly reversible, with periods of restricted muscle use, resulting in loss of muscle fibre mass (atrophy).*

Effect of aerobic training on skeletal muscle seen in cross section.
All fibres increase in thickness. Mitochondria, blood capillaries, glycogen stores, fat vacuoles all increase.

**Untrained**                                                    **Trained**

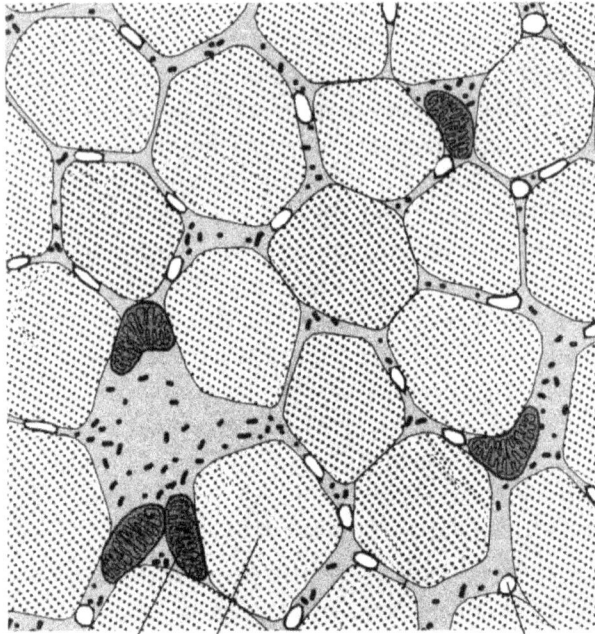 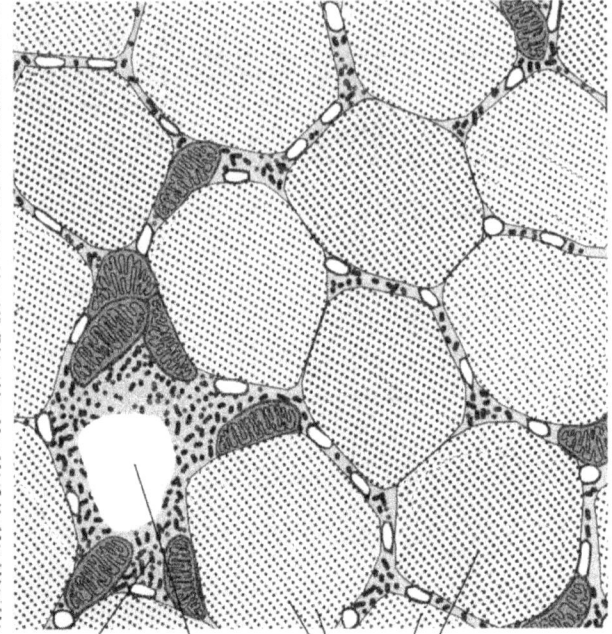

mitochondrion    myofibrils; made of many filaments    sarcoplasmic    glycogen    fat vacuole    individual filaments    ⌐1.0μ⌐
                 and in turn many myofibrils make      reticulum       granules
                 up a single muscle fibre

**Table 1.2:** *Summary of differences between fast and slow twitch muscle fibres.*

| Characteristic | ST Type I *(SO)* | FT Type II | |
| --- | --- | --- | --- |
| | | Type IIa *(FOG)* | Type IIb *(FG)* |
| Contraction time | Slow *(110 ms)* | Fast *(50 ms)* | Fast *(50 ms)* |
| Aerobic capacity | High | High | Low |
| Anaerobic capacity | Low | High | High |
| Force of motor unit contraction | Low | High | High |
| Mitochondrial number | High | High | Low |
| Myoglobin content | High | Medium | Low |
| Phosphocreatine store | Low | High | High |
| Glycogen store | Medium | High | High |
| Elasticity | Low | High | High |
| Activity level of myosin ATP-ase | Low | High | High |
| Amount of sarcoplasmic reticulum and rate of calcium release | Low | High | High |
| Muscle fibre | Small diameter | Large | Large |
| Recruitment threshold | Low | High | High |

## 1.5 MOTOR UNITS

Skeletal muscles are stimulated by motor nerves. Each motor nerve contains many motor neurones or nerve cells, but there are far more muscle fibres than there are motor neurones, therefore each motor neurone must supply more than one muscle fibre. A motor neurone and the muscle fibres it stimulates, act as a unit, and are together known as a **motor unit**.

The number of muscle fibres stimulated by a single motor neurone can vary from very few, giving fine control, for example in the muscles that move the eye in its socket; to very many, for example, in the quadriceps muscle at the front of the thigh, where control is correspondingly less fine.

### 1.5.1 MOTOR UNIT TYPES

Muscle fibres in a particular motor unit are all of the same type. Indeed the fibre type of a motor unit is mainly *(but not entirely)* determined by the nature of the motor neurone supplying it. This is illustrated by the fact that the only clear case of one type of muscle fibre being converted to another, is when a motor neurone supplying fast Type II muscle fibres is switched surgically to supply slow Type I muscle fibres, and vice versa. Fast fibres are in fact converted to slow fibres more easily than the reverse. Because of this interconvertibility it is perhaps more correct to refer to 'slow' and 'fast' motor units rather than muscle fibres. There are less fibres in a 'slow' Type I motor unit (up to about 200) than in a 'fast' Type II motor unit (up to a thousand or more), resulting in Type II motor units generating greater force.

The motor unit is a functional unit, and the neural contribution to strength of contraction and involvement in fatigue should be remembered.

**Table 1.4:** *Comparison of neurone types.*

| Motor Neurone | Supplying Muscle Fibre Type | |
| --- | --- | --- |
| | I *(ST)* | II *(FT)* |
| Cross section area of nerve fibre | Smaller | Larger |
| Conduction Velocity | Slow | Fast |

## 1.6 NERVOUS STIMULATION OF MUSCLE FIBRE

At the point of contact with the muscle fibre, the motor neurone terminates in a motor end plate, otherwise known as the neuromuscular or myoneural junction. This is where the nerve impulse stimulates the muscle fibre.

**Figure 1.6:** *A neuromuscular junction as seen in section.*

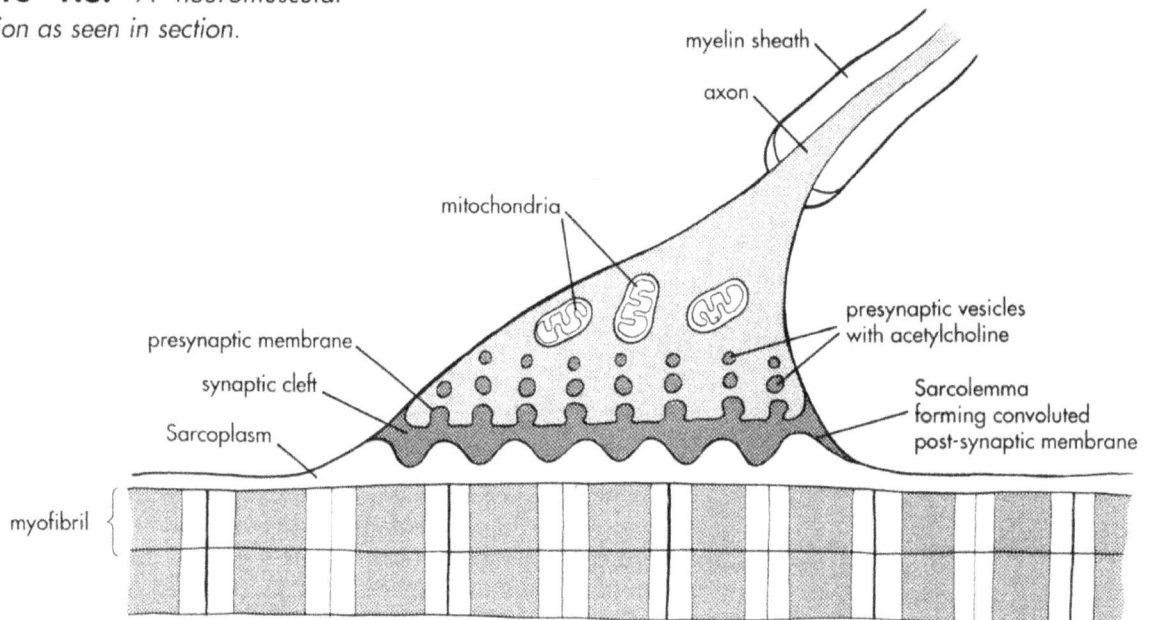

When a nerve impulse arrives at the motor end plate, the presynaptic vesicles release the neurotransmitter substance acetylcholine. This diffuses across the gap *(known as the synaptic cleft)* between the motor end plate and the post-synaptic membrane of the muscle, and initiates a wave of electrical activity *(an excitatory post synaptic action potential (EPSP))* in the post synaptic membrane. If the EPSP exceeds a certain threshold it will initiate a contraction of the muscle fibres in the motor unit. If it does not, but the nerve impulses are repeated at a high enough frequency, the successive discharges from the presynaptic membrane add up or summate, resulting in an increased EPSP which is effective. Generally speaking slow twitch muscle fibres are less excitable than fast twitch muscle fibres.

After the action is completed the enzyme cholinesterase breaks down the acetylcholine, thus clearing the gap in readiness for the arrival of another impulse at the motor end plate.

### 1.6.1 TRANSMISSION WITHIN THE MUSCLE FIBRE

The action potential spreads rapidly over the surface membrane of the fibre and throughout all the myofibrils via in-turnings of the membrane, known as T tubules. The T tubules are in close association with the sarcoplasmic reticulum around each myofibril.

**Figure 1.7:** *Sarcoplasmic reticulum.*

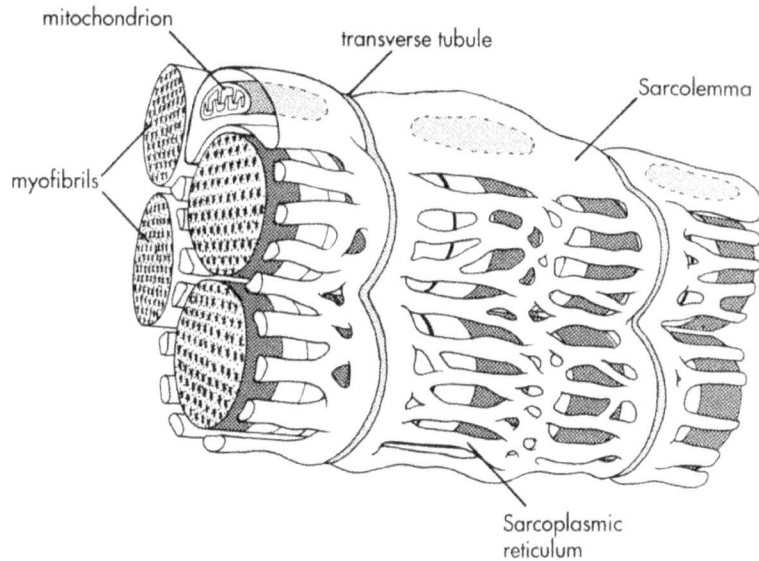

The arrival of the action potential triggers the release of calcium ions from the sarcoplasmic reticulum vesicles. This event initiates the contraction process, as previously described *(Section1.3.2)*.

As a result of the arrangement of the T tubules, which penetrate all parts of the fibre, all the myofibrils contract together. Also as each muscle fibre in a motor unit receives the motor nerve impulse at the same time, all the fibres in a motor unit contract together.

## 1.6.2 RESPONSE OF MUSCLE TO STIMULATION

Nerve impulses are all of the same strength or magnitude, and travel along the motor nerve fibre or axon either at full strength or not at all, obeying what is known as the 'all or nothing law'. Under certain conditions muscle fibres and therefore motor units, also obey this law and either contract completely or not at all. However, the degree of muscle fibre contraction can decrease as the fibres fatigue, or it may even increase as a result of a previous stimulation.

Variation in the strength of the muscle response as a whole is achieved in different ways. One method of varying the strength of the muscle contraction, is achieved by what is known as **spatial summation** or **multiple motor unit summation**, in which the number and the size of the motor units recruited throughout the muscle as a whole are varied. A staggered spread of activation of the motor units throughout the muscle, also enables sustained contractions to be maintained, as some motor units are contracting whilst others are relaxing. In this way the use of ATP, and the associated fatigue, is spread throughout the muscle, instead of being confined to certain units.

Another method of varying the strength of the muscle contraction involves the frequency of motor nerve impulses. If the frequency of motor nerve impulses is such that the next impulse reaches a muscle before it has completely relaxed after the previous contraction, it starts the second contraction from a higher force level, so that the resulting force is greater than that from a single stimulus. This adding together or **wave summation** of contractions can lead, if the impulses arrive so fast that there is no time for any relaxation, to a complete fusion of the contractions, which endures as long as the stimuli persist, or until the muscle fatigues.

---

### ❖ *Training and Performance Applications*

*A type of complete fusion of contractions experienced by many sportspersons is muscle cramps. These can occur during exercise, usually after strenuous effort in hot conditions, or even whilst inactive. Although correct conditioning in a controlled training programme can reduce the risk of muscle cramps, there is no clear explanation of why these cramps occur. They are generally short lasting and cannot be reliably reproduced under controlled conditions, so it is not possible to investigate all the factors that might be involved in causing them. Suggested explanations include imbalances in the normal amounts of sodium, potassium, and chloride ions on either side of the muscle membranes, and/or a failure to withdraw calcium ions back into the sarcoplasmic reticulum at the end of the stimulation of the muscle for some reason.*

---

## 1.7 MUSCLE FATIGUE

Although it is obvious that muscles tire as a result of strenuous use, such are the complexities of muscle physiology that the causes of muscle fatigue are not easily identified. Contributing factors include the following.

**Interruption of neuromuscular events.**
Depletion of the neurotransmitter substance acetylcholine, which is released by the motor neurones at the neuromuscular junction, reduces the stimulation of the muscles. This would appear to be more common with fast twitch motor units, and could help explain why they fatigue more rapidly compared with the slow twitch motor units.

Once the muscle membranes are stimulated, the impulse may not be propagated throughout the muscle due to an imbalance between sodium and potassium ions across the muscle membranes.

Once the impulse reaches the muscle fibres, there may be reduced calcium ion release, resulting in reduced cross bridge formation. This reduced calcium ion release may be due to either temporary T-tubule damage, or the inhibiting effect of accumulating lactic acid.

**Depletion of the sources of energy.**
Fatigue will result as it becomes increasingly difficult to replenish supplies of ATP *(Chapter 2 Section 2.4, p 43)*. If phosphocreatine stores are depleted, as they are when running the 100 and 200 metres for example, maximum power cannot be maintained.

Depletion of muscle and liver glycogen produces a variety of effects, all of which contribute to fatigue. Glycogen is a readily available source of energy under both aerobic and anaerobic conditions, and as the stores are used up, the same level of activity cannot be maintained. Also, fatty acids from the fat stores cannot be fully oxidised to release all their energy if there is insufficient breakdown of carbohydrates at the same time. Under these conditions poisonous by-products accumulate, leading to the rapid onset of fatigue.

The depletion of a substance called carnitine, which aids the transport of the fatty acids into the mitochondria, where they are oxidised, also contributes to fatigue.

**Alterations in the concentration of the body fluids.**
Loss of water from the body, and therefore also the muscles, will disturb ion balance and lead to disruption of normal functioning as described before. Water loss can also contribute to an increase in body temperature due to a decrease in sweating as dehydration sets in *(Chapter 2 Section 2.8, p 54)*.

**Increase in body temperature.**
Active muscle rapidly heats up by about 2-3°C. Generally this is beneficial as it speeds up the rate of enzyme catalysed reactions involved in muscle contraction, and increases the amount of oxygen liberated from the oxyhaemoglobin in the muscle capillaries. However, an increase of more than 2-3°C can disrupt the balanced biochemistry of muscle metabolism, quickly leading to fatigue.

**Insufficient blood supply.**
An insufficient blood flow to the muscles will increase most of these factors contributing to fatigue, particularly by not delivering sufficient oxygen to maintain aerobic respiration. One cause of this in prolonged exercise in the vertical position is postural hypotension, where blood accumulates in the veins of the lower parts of the legs, reducing cardiac output as a result of reduced return of blood to the heart.

As the oxidation of fat requires more oxygen than the oxidation of glucose, the increasing use of fat during exercise increases the problem of oxygen shortage. With insufficient oxygen, respiration becomes anaerobic and lactic acid accumulates. The increased acidity *(decreased pH)* has a complex of effects, all of which contribute to fatigue. It inhibits resynthesis of creatine phosphate, and interferes with actin/myosin cross bridge formation, possibly by interfering with the binding of calcium ions to troponin. It also inhibits enzymes, especially those involved in glycolysis, and decreases the amount of energy released by the hydrolysis of ATP.

### ❖ *Training and Performance Applications*

*An artificial method of trying to overcome the fatigue effects of increasing acidity, is to consume sodium hydrogencarbonate (sodium bicarbonate), an alkali (antacid), just before the event. This technique is referred to as 'Soda Loading', and the increased alkalinity (increased pH) of the blood increases the capacity to neutralise the lactic acid. The timing is critical as it may cause nausea and diarrhoea, and the increase in alkalinity is itself buffered and removed from the body. The ingestion of sodium citrate also increases the plasma hydrogencarbonate levels, and avoids any gastro-intestinal problems.*

*Events causing localised muscle fatigue can also exert an effect on the brain, causing what is known as central fatigue. For example depletion of blood glucose can affect the brain, because glucose is its sole respiratory substrate.*

*It is also suggested that changes in the concentration of various amino acids in the blood as a result of exercise may affect the brain and therefore the perceived effort. The brain detects these changes and exerts an inhibitory effect on the motor system, in an attempt to reduce the level of fatigue experienced. The extent to which conscious 'will power' can overcome this sub-conscious inhibitory effect of the brain on the motor system, and maintain maximum effort is not measurable. Many elite athletes claim that superior will power gains them victory, but this is just a personal conceit. It could be quite possible that the person coming last is trying even harder than the winner. Certainly a strong will power would ensure the best training preparation, and could over ride the weakening effect of competition 'nerves', but on the day, winning is much more likely to be 'all in the body' rather than 'all in the mind'.*

## 1.8 PROPRIOCEPTORS

Proprioceptors are sense organs in muscles, tendons and joints, which provide information about the position and movement of the body in what is known as kinaesthenic feedback. There are several types of proprioceptors located in different parts of the skeletal system.

## 1.8.1 MUSCLE SPINDLE APPARATUS

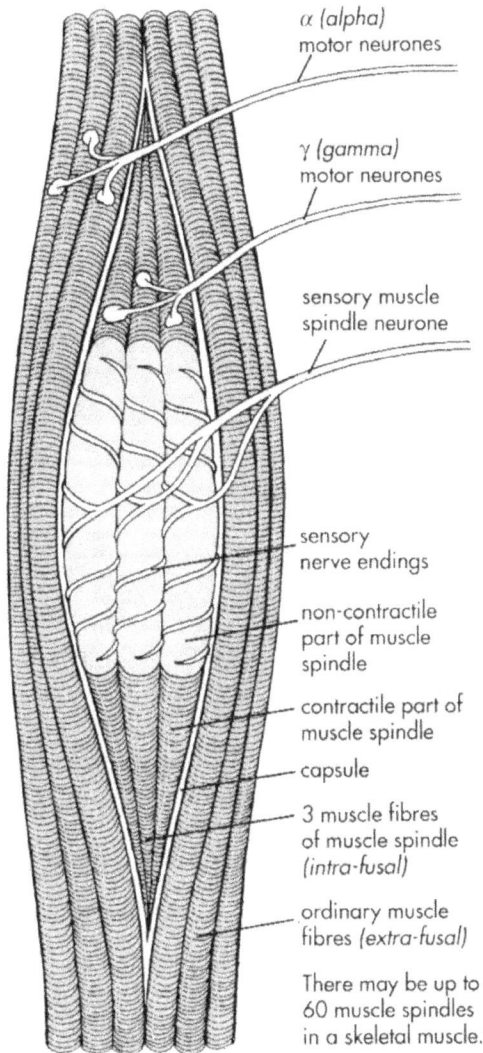

Each muscle spindle apparatus consists of several modified muscle fibres *(intrafusal fibres)* bound in a fine connective tissue sheath. They are supplied with both motor and sensory neurones.

Muscle spindles are involved in the detection of the state of contraction/extension of a muscle, the anticipation of muscle loading, and in the actual fine control of muscle contraction.

**Figure 1.8:** *Muscle Spindle Apparatus structure and function.*

**Muscle spindle apparatus**

α (alpha) motor neurones

γ (gamma) motor neurones

sensory muscle spindle neurone

sensory nerve endings

non-contractile part of muscle spindle

contractile part of muscle spindle

capsule

3 muscle fibres of muscle spindle (intra-fusal)

ordinary muscle fibres (extra-fusal)

There may be up to 60 muscle spindles in a skeletal muscle.

**Role in detection of muscle contraction**

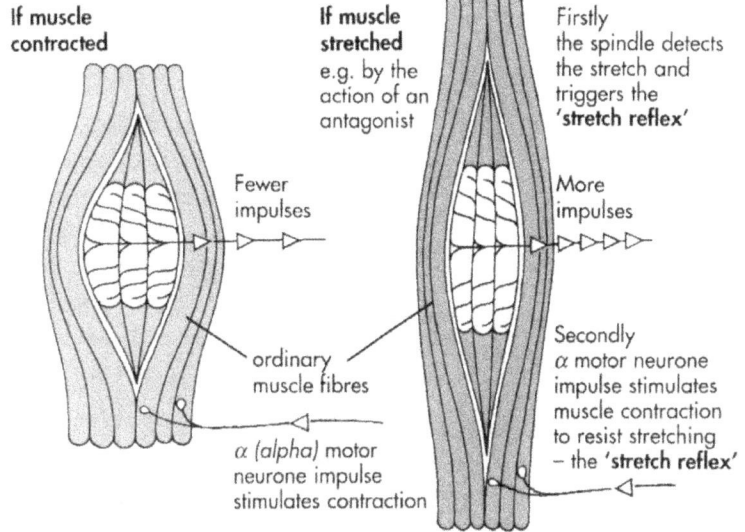

**If muscle contracted**

Fewer impulses

ordinary muscle fibres

α (alpha) motor neurone impulse stimulates contraction

**If muscle stretched** e.g. by the action of an antagonist

More impulses

Firstly the spindle detects the stretch and triggers the **'stretch reflex'**

Secondly α motor neurone impulse stimulates muscle contraction to resist stretching – the **'stretch reflex'**

**Gamma loop control in fine muscle action**

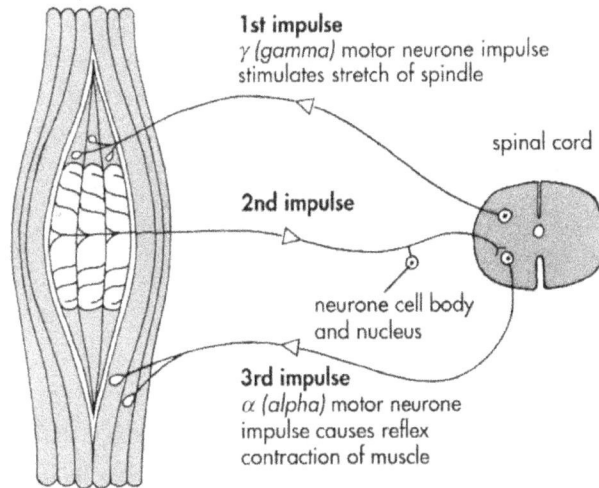

**1st impulse** γ (gamma) motor neurone impulse stimulates stretch of spindle

spinal cord

**2nd impulse**

neurone cell body and nucleus

**3rd impulse** α (alpha) motor neurone impulse causes reflex contraction of muscle

## 1.8.2 DETECTION OF STATE OF CONTRACTION

Contraction, or stretching, of a muscle alters the tension exerted on the muscle spindle apparatus. This is detected by the sensory nerve endings in the central part of the spindle. Sensory feedback of the state of contraction, or stretch of the muscle, is thus relayed to the central nervous system.

### ❖ Training and Performance Applications

*Stretching plays an important role in training and just before competition. If a muscle is suddenly stretched, as are the hamstrings in a bouncing style of toe touching exercise, the* **stretch reflex** *is activated. The muscle spindles detect the stretch and trigger an opposing reflex contraction of the muscle being stretched. Such 'stretching' exercises are therefore counter-productive, as the muscle contracts to protect itself. The strength of the stretch reflex depends on the speed of the stretch. The faster the stretch, the stronger the reflex contraction of the muscle being stretched. Sustained stretching leads to a reduction in the signals from the muscle spindles, and the stretch reflex is decreased in strength in a process known as* **habituation***.*

*The connective tissue that binds muscle fibres together into bundles, and which is continuous with the tendons that connect the muscles to bone, is capable of being stretched to a certain degree, and the muscle fibres themselves grow in length in response to regular stretching. Both these adaptations increase the range of movement possible about a joint. The range of movement about a joint is also affected by the joint structure and amount of sub-cutaneous fat.*

### 1.8.3 ANTICIPATION OF MUSCLE LOADING

The muscle spindle can be set at a prescribed firing tension, feeding back sensory information when stretched beyond that certain tension. This pre-setting of muscle spindles is known as **gamma bias** and allows the detection of the stretching of a muscle at different degrees of its contraction. When wishing to maintain a muscle contraction of a certain tension, the gamma bias can be reset so that stretching of this now partially contracted muscle can be detected. An example is seen when one prepares to lift a load, the weight of which is anticipated as being 'heavy'.

### ❖ Training and Performance Applications

*Anticipating the force of muscle action required under varying circumstances, plays a major role in sport participation. Consider for example the fine control exerted in racquet sports, such as tennis and squash. The ball approaches at a wide variety of speeds, and yet the return shot is invariably weighted correctly within very fine limits. In weight lifting, misjudgements leading to injury serve to highlight the protective function of correct anticipation. Although this is basically a reflex action, there is an element of conscious input into the process.*

## 1.8.4 FINE CONTROL OF MUSCLE ACTION

Stimulation of skeletal muscle can either be direct via the **alpha** motor neurones which innervate the muscle fibres, or indirectly via the muscle spindles.

With indirect control via the muscle spindles, the **gamma** motor neurones stimulate the contractile ends of the intrafusal fibres within the muscle spindle, thus stretching the sensory stretch receptor endings in the centre. This stretching causes reflex contraction of the extra-fusal or normal muscle fibres via the alpha motor neurones.

Stimulation of muscle contraction via the muscle spindles takes longer than direct alpha motor fibre stimulation, as not only is the nerve pathway via the muscle spindle longer, but the gamma neurones have a smaller diameter and therefore conduct impulses more slowly than the alpha neurones. Activation via the gamma motor system can give finer control than direct alpha motor control, but if the gamma motor system is used whilst the muscle fibres are contracting under alpha motor stimulation *(alpha-gamma co-activation)*, then the muscle spindle reflex reinforces the muscle contraction, producing greater force.

Skeletal muscles typically occur in pairs, for example the biceps and triceps of the upper arm. The stimulation of a muscle to contract typically leads to the inhibition of the muscle that works in opposition to it *(reciprocal inhibition)*, so that it cannot contract at the same time and oppose the intended action, which would occur as a result of the stretch reflex. So in this example, if the biceps contracts, the triceps relaxes; and vice versa.

❖ *Training and Performance Applications*

*Muscle co-ordination and reactions are improved by a proper 'warm-up' period, so that the reciprocal inhibition between antagonistic muscles operates more efficiently. If performers of explosive power events, such as sprinters, field event athletes, and games players, do not carry out such a 'warm-up' correctly (not to be confused with simple static stretching), then they are prone to muscle 'tears' and 'pulls', caused as a result of one of a pair of antagonistic muscles not relaxing quickly enough as the other one contracts. The classic example of this is the 'hamstring' injuries suffered by sprinters.*

### 1.8.5 OTHER SENSORY RECEPTORS

**Golgi tendon organs.**

These occur at the junction of a tendon and its muscle. They detect the degree of stretch of the tendon. When stretched, they trigger the reflex inhibition of the contracting muscle that is causing the tendon to be stretched, and the reflex contraction of the antagonist *(reciprocal activation)*. This prevents damage to the muscle and tendon. Golgi tendon organs are less sensitive than muscle spindles, and only come into operation at high, potentially damaging tensions.

---

❖ *Training and Performance Applications*

*It is thought that stretching may relieve muscular cramp as a result of the Golgi tendon organs sending inhibitory impulses to the muscle involved, causing it to relax. These protective reflexes of the tendon organs also operate more efficiently after a proper 'warm-up' period. If a muscle is stretched for longer than about six seconds, the Golgi tendon organs are activated, resulting in relaxation of the stretched muscle; this is known as the inverse stretch reflex as it opposes the stretch reflex.*

*If a limb is moved passively (i.e. with the assistance of a partner) to the limits of its range of movement and the muscles then contracted isometrically (without shortening) for about six seconds in that position, and then the muscles are relaxed; the limit of extension will be increased as a result of proprioceptive neuromuscular facilitation (PNF). That is the proprioceptors (muscle spindle apparatus and Golgi tendon organs), are facilitating (helping) stretching by means of neuromuscular events. In other words the passive stretch and the contracted muscle acting on the tendon result in the Golgi tendon organ causing further muscle relaxation, which helps further stretching; at the same time the muscle spindle apparatuses are habituated and any reflex contraction caused by the stretching is decreased.*

---

**Joint receptors.**

These are sensory nerve endings found in the joint capsules and ligaments which detect movement and position of the joint.

**Pain receptors.**

Pain receptors are scattered throughout the muscle fibres, their connective tissue and blood vessels. Stimulation of these receptors is experienced as pain.

# 1.9 ANATOMY OF MOVEMENT

## 1.9.1 TYPES OF SKELETAL MUSCLES.

Muscle fibres are bound by connective tissue into bundles. Within these bundles the fibres are arranged parallel to the longitudinal axis. The bundles, however, are arranged within muscles in a variety of patterns. The pattern, along with the particular attachment of the muscle to the skeleton, determines the force and range of movement that can be produced

**Fusiform** muscles have comparatively few long bundles, arranged parallel with the longitudinal axis of the muscles, running into flat, strap-like tendons at each end. This arrangement allows maximum shortening, and they are attached in positions that can result in rapid and wide ranging movements. An example is seen in the biceps brachii which flexes the arm.

**Pennate** muscles are broad and flat with comparatively more, shorter bundles radiating out from a single tendon that extends the length of the muscle. Muscle fibres can contract to just over half of their resting length. Therefore the shorter fibres of pennate muscles cannot contract by as much as those of fusiform muscles, but as there are more of them, the pennate muscle can contract more forcefully. Pennate muscles are attached in positions that can result in powerful movements, for example the rectus femoris of the thigh, which straightens the leg at the knee.

## 1.9.2 TYPES OF MUSCLE ACTION

Muscles act in different ways with regard to changes in length and tension; and the interaction of muscular force, the load or resistance, and the action of the bones as levers, determines the outcome.

Although the sliding filament mechanism must be active for a muscle to develop a force, the sarcomeres (and the muscle as a whole) do not always shorten. If the force that they generate is greater than the resistance, then they will shorten. If the force they generate cannot overcome the resistance they will not shorten; and if they are overcome by the resistance they will actually lengthen under tension.

### Dynamic actions *(Isotonic)*

The most common type, resulting in movement at the joints. There are two types; concentric and eccentric.

### Concentric actions

These are perhaps the most familiar types of contraction, in which the whole muscle shortens as it develops tension.

This type of contraction is seen for example in the biceps muscle of the arm when 'curling a weight'.

### Eccentric action.

Here the muscle develops its active tension whilst it is being lengthened by stretching. One example of this is seen in the quadriceps muscle at the front of the thigh when walking or running down steps or down hill. Another is seen in the biceps of the arms as they control the lowering of a weight after a curl.

### Static actions (Isometric)

These occur when the muscle is prevented from shortening whilst it is generating tension, as for example when straining against an immovable resistance. Even though the bones involved may not be moving, initially the sliding filament mechanism operates normally as the 'slack' in the elastic soft tissues is taken up, but once the steady state is achieved the cross bridges hold the filaments steady in one position.

Although classified separately in this way, isometric and isotonic contractions are both involved in the typical muscle contraction that results in movement; isometric contraction occurs initially before sufficient force is generated to actually start the movement, and isotonic contraction occurs once the movement is started.

**Figure1.11:** *Major body axes and planes. To help the understanding of movement of segments of the body with respect to one another, imaginary reference planes are used. Within each plane an axis can be identified, usually in association with a particular joint, about which movement takes place.*

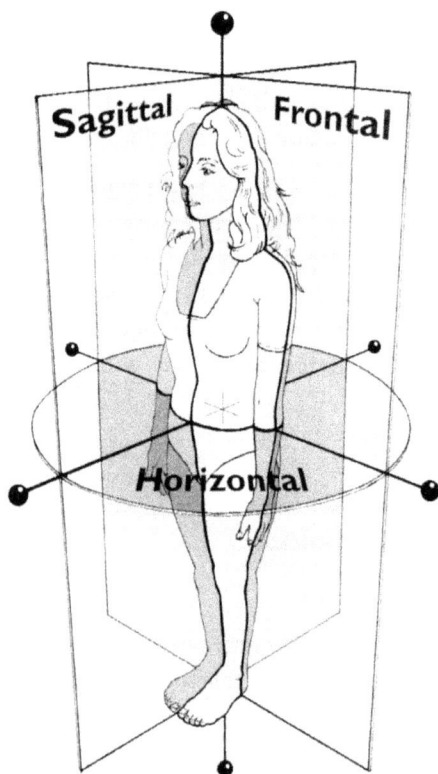

Sagittal    Frontal

Horizontal

---

### ❖ *Training and Performance Applications*

*Isokinetic* contractions occur when the muscle shortens at constant speed, and exerts maximum tension over the full range of movement at all joint angles, thus stimulating maximum gains over the whole range of joint movement. This is only fully achieved with specially designed weight training apparatus, which automatically adjusts the resistance to match the muscle force over the full range of movement. However, this constant movement over the whole range of joint movement is rarely met with in sport. The closest natural example of such isokinetic contraction in sport, is seen in the freestyle swimming armstroke.

*Resistance or weight machines are safe to use, convenient, and popular in fitness centres; however free weights are preferred by weight lifters as they involve more muscle groups in more dynamic action.*

*A training method known as **plyometrics** uses both concentric and eccentric contractions. In plyometrics, maximum concentric effort is made immediately following an eccentric phase, in which the muscles have been stretched whilst developing tension. Muscles contract with greater force as a result of the stretch reflex being stimulated to help recruit extra motor units. This is further reinforced by passive elastic recoil of the muscle, when the muscle is extended beyond its resting length, before being stimulated to contract. In this way muscles generate what can be referred to as explosive power. This is achieved for example by bounding and hopping, and is widely used by athletes wishing to increase their leg power.*

### 1.9.3 MOVEMENT IN RELATION TO THE MAJOR BODY PLANES AND AXES

**Figure 1.12:** *Movement in relation to the major body planes and axes.*

Voluntary muscles exert their action via the skeleton, which acts as a series of levers, articulated at joints, which allow movement in all planes.

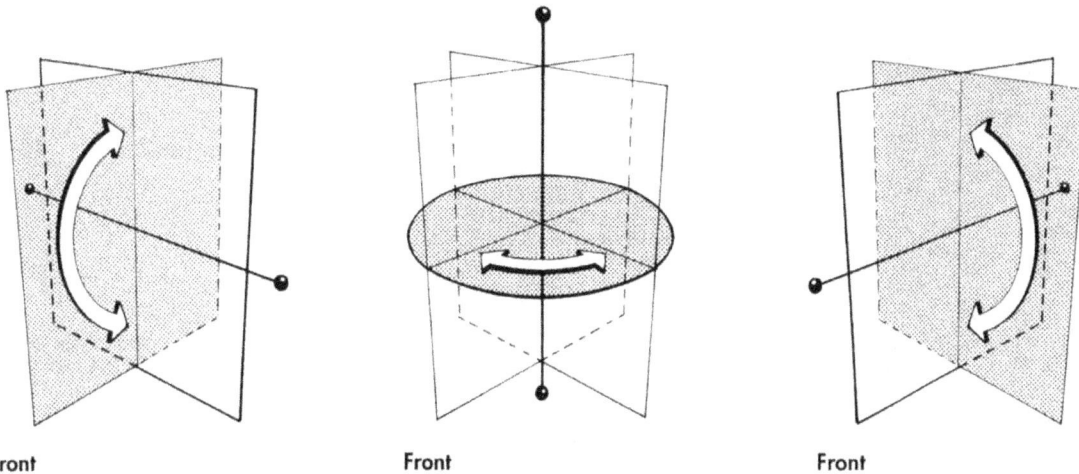

Front

Front

Front

**(i)** *Side to side (transverse) axis, involved in movement in the sagittal plane known as **flexion** and **extension**, e.g. bending (flexion) and straightening (extension) the arm at the elbow.*

**(ii)** *Vertical axis, involved in movement in a transverse plane known as **rotation**, e.g. rotating the head from side to side.*

**(iii)** *Front to back axis, involved in movement in a frontal (coronal) plane known as **abduction** and **adduction**, e.g. raising the arm sideways at the shoulder (abduction), and lowering it (adduction).*

### 1.9.4 MUSCLE ARRANGEMENTS

The muscles act in groups rather than singly, with most arranged in opposed or **antagonistic pairs** at joints, for example flexors and extensors. The muscle that causes the required action is called the prime mover or **agonist** *(protagonist)*, for example the biceps brachii flexing the arm at the elbow. At the same time the **antagonist** muscle, in this case the triceps brachii, is inhibited and relaxes, allowing the flexion to occur. If the arm is extended at the elbow, then the roles of agonist and antagonist are reversed, with the triceps brachii being the agonist, and the biceps brachii being the antagonist.

Muscles known as **fixators** or **stabilisers** hold or fix joints in a stable position, as for example when keeping the wrist stiff during a biceps curl with weights.

Others known as **synergists** prevent any unwanted movement and help the prime mover to function more efficiently.

**Figure 1.13:** *The Human skeleton*

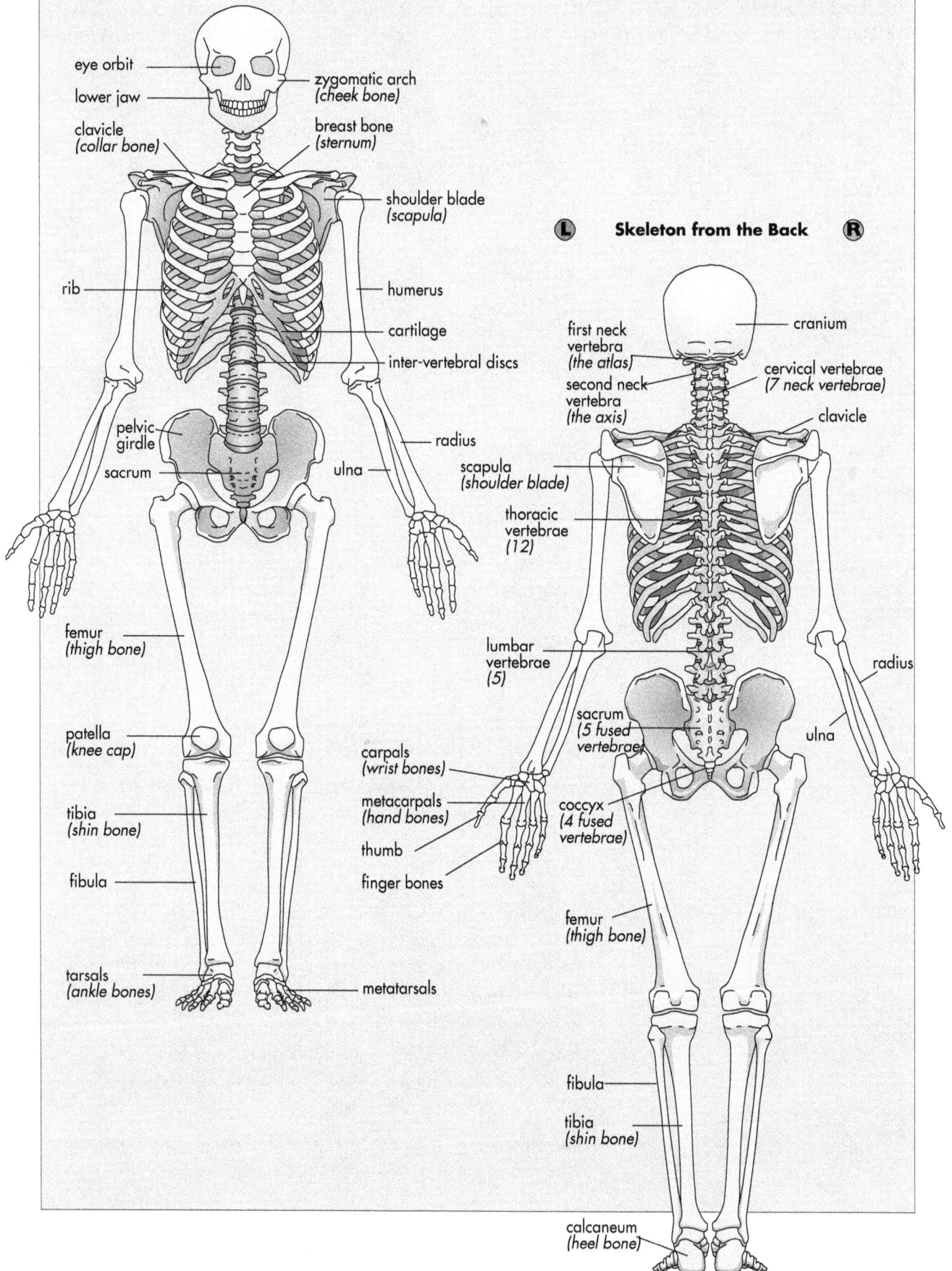

**®** **Skeleton from the Front** **L**

eye orbit

lower jaw

clavicle
(collar bone)

rib

pelvic girdle

sacrum

femur
(thigh bone)

patella
(knee cap)

tibia
(shin bone)

fibula

tarsals
(ankle bones)

zygomatic arch
(cheek bone)

breast bone
(sternum)

shoulder blade
(scapula)

humerus

cartilage

inter-vertebral discs

radius

ulna

carpals
(wrist bones)

metacarpals
(hand bones)

thumb

finger bones

metatarsals

**L** **Skeleton from the Back** **®**

first neck
vertebra
(the atlas)

second neck
vertebra
(the axis)

scapula
(shoulder blade)

thoracic
vertebrae
(12)

lumbar
vertebrae
(5)

sacrum
(5 fused
vertebrae)

coccyx
(4 fused
vertebrae)

femur
(thigh bone)

fibula

tibia
(shin bone)

cranium

cervical vertebrae
(7 neck vertebrae)

clavicle

radius

ulna

calcaneum
(heel bone)

**Figure 1.14:** *Superficial muscles of the human body. Not all muscles are shown and those that are have been simplified*

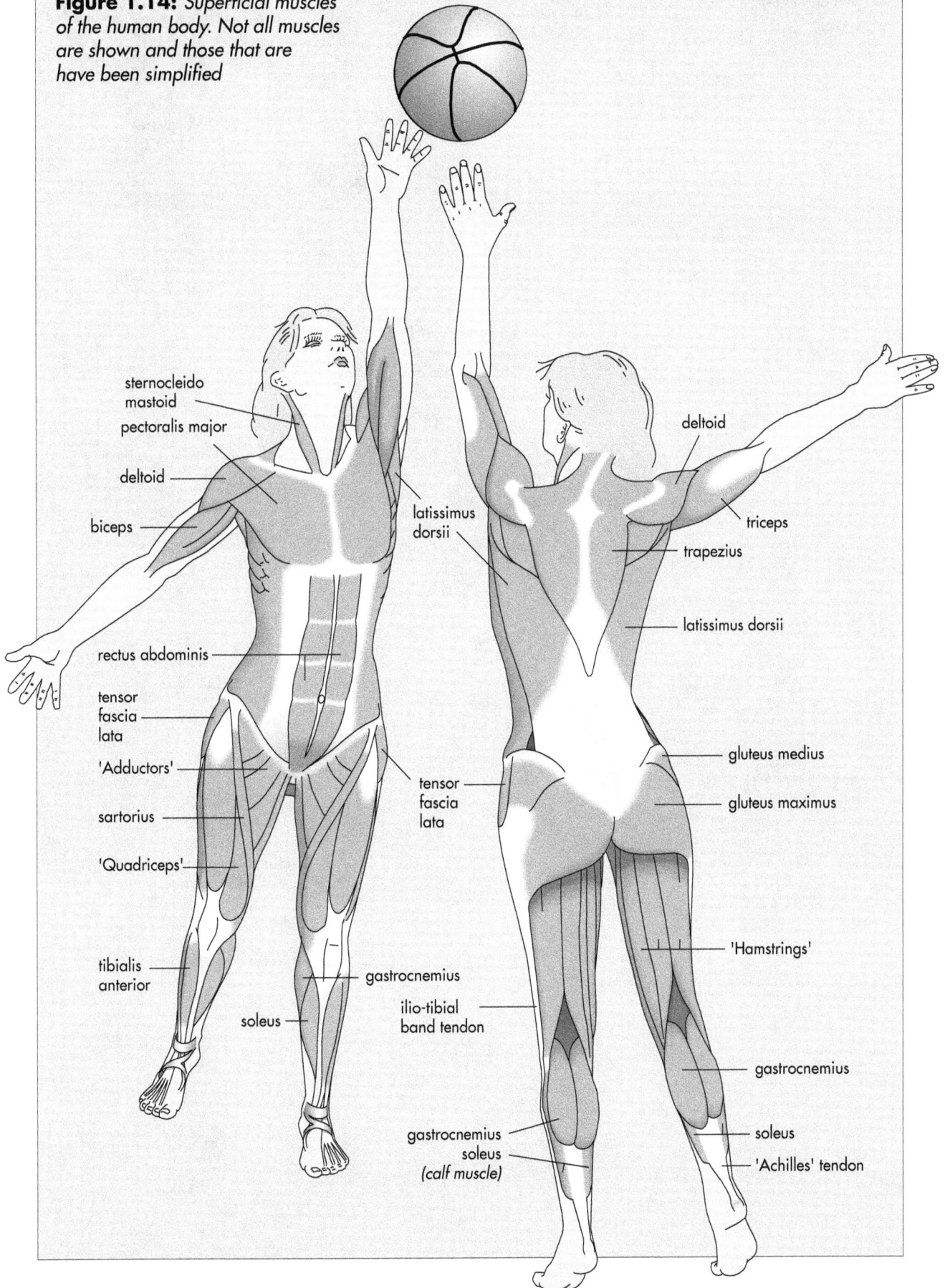

sternocleido mastoid

pectoralis major

deltoid

biceps

latissimus dorsii

deltoid

triceps

trapezius

latissimus dorsii

rectus abdominis

tensor fascia lata

'Adductors'

sartorius

'Quadriceps'

tensor fascia lata

gluteus medius

gluteus maximus

tibialis anterior

gastrocnemius

'Hamstrings'

soleus

ilio-tibial band tendon

gastrocnemius

soleus

'Achilles' tendon

gastrocnemius
soleus
*(calf muscle)*

**Figure 1.15:** *Main types of movable joint*

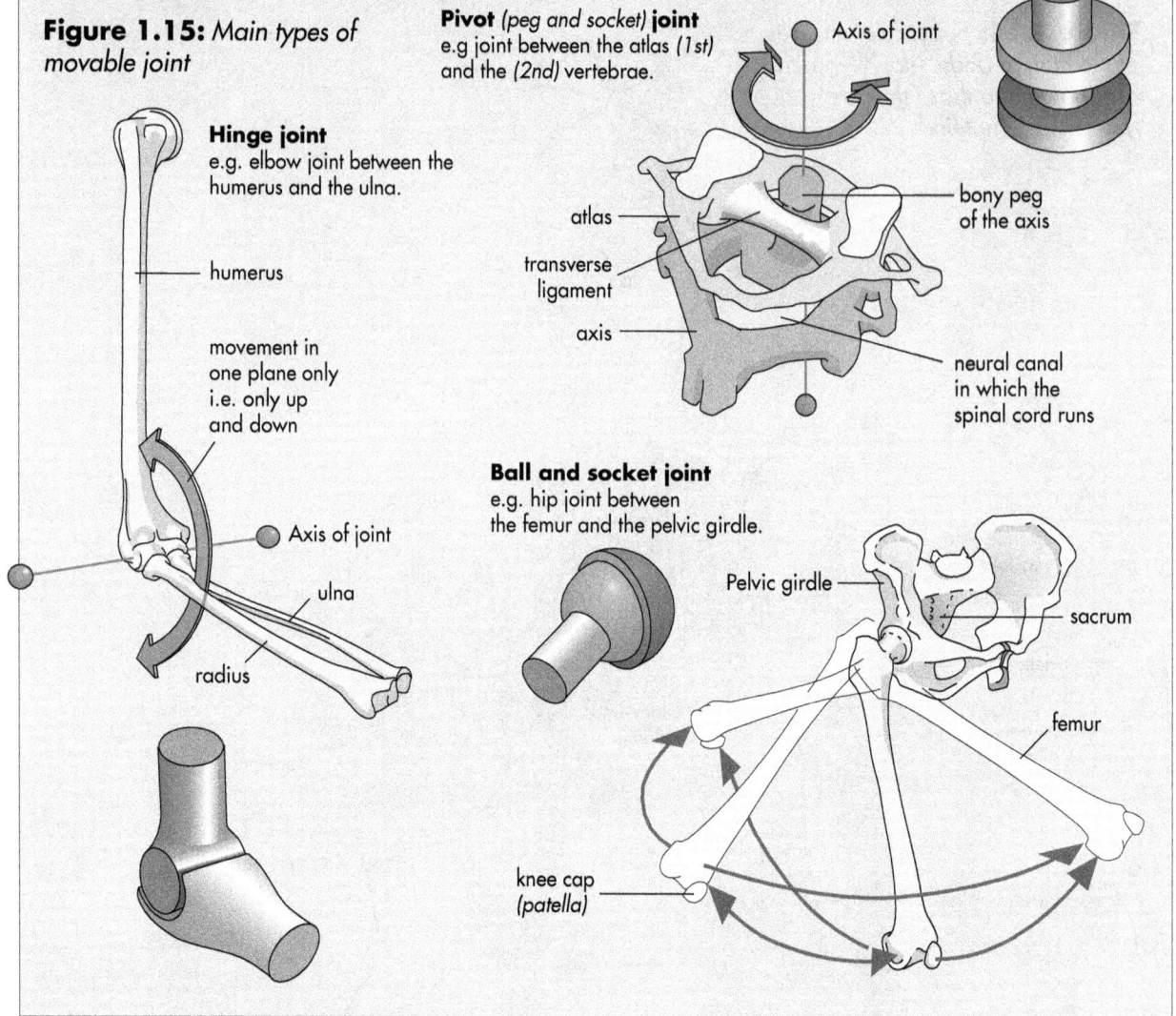

**Hinge joint**
e.g. elbow joint between the humerus and the ulna.

humerus

movement in one plane only i.e. only up and down

Axis of joint

ulna

radius

**Pivot** *(peg and socket)* **joint**
e.g joint between the atlas *(1st)* and the *(2nd)* vertebrae.

Axis of joint

atlas

transverse ligament

axis

bony peg of the axis

neural canal in which the spinal cord runs

**Ball and socket joint**
e.g. hip joint between the femur and the pelvic girdle.

Pelvic girdle

sacrum

femur

knee cap *(patella)*

**Synovial joint structure**
as seen in section of the hip joint.

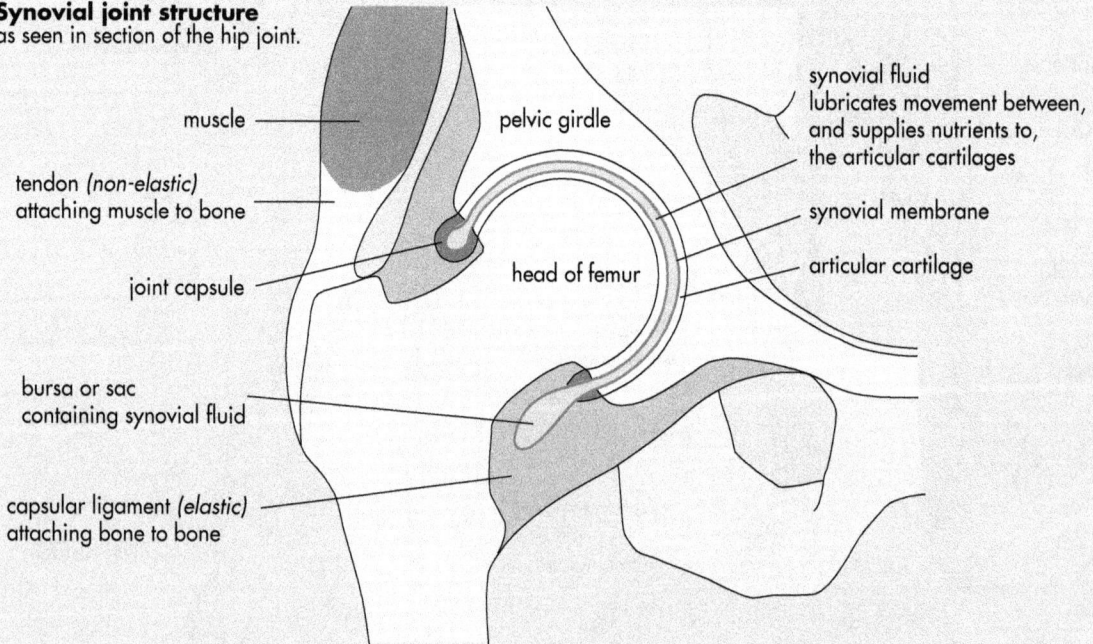

muscle

pelvic girdle

tendon *(non-elastic)* attaching muscle to bone

joint capsule

head of femur

synovial fluid lubricates movement between, and supplies nutrients to, the articular cartilages

synovial membrane

articular cartilage

bursa or sac containing synovial fluid

capsular ligament *(elastic)* attaching bone to bone

**Front View**

**Major Muscles of the Trunk**

Ⓡ    Ⓛ

rectus abdominis muscles set in tendinous sheet (attached to sternum, rib cage and pelvic girdle)

**Back View**

Ⓛ    Ⓡ

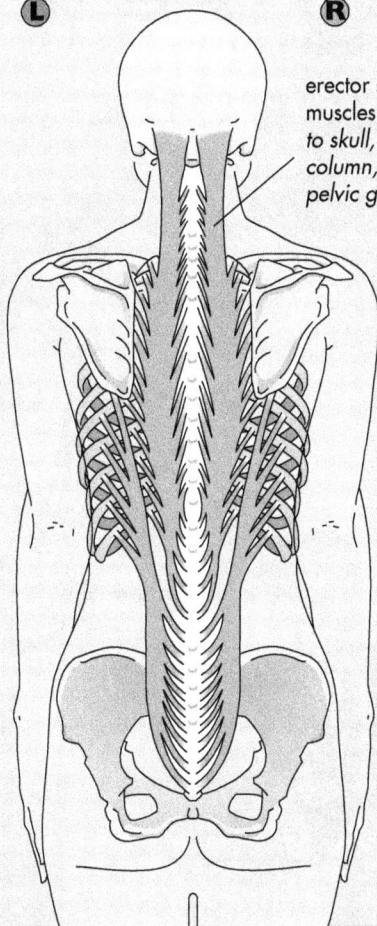

erector spinae muscles (attached to skull, vertebral column, ribs, and pelvic girdle)

**Lateral Flexion**

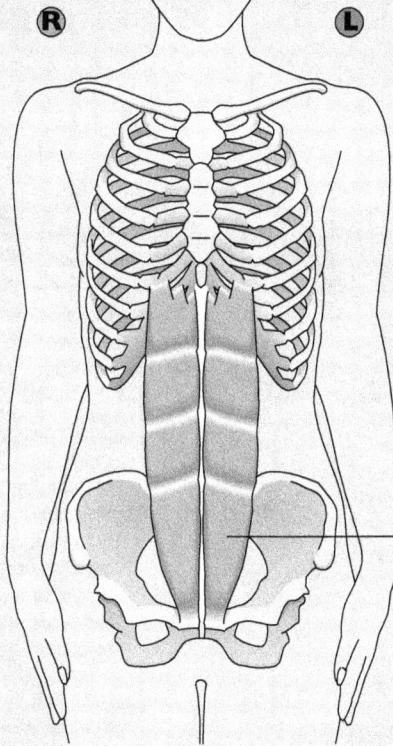

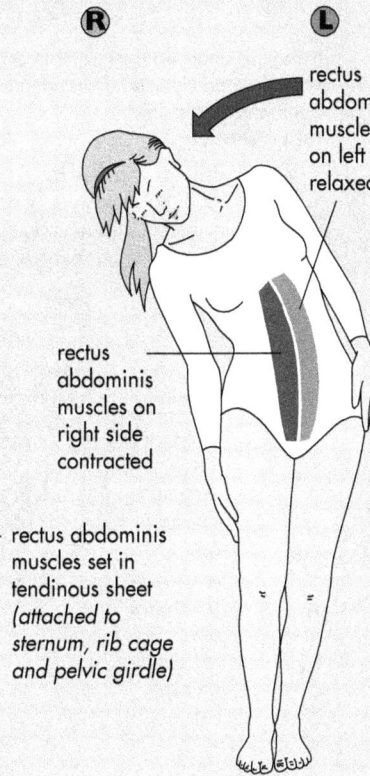

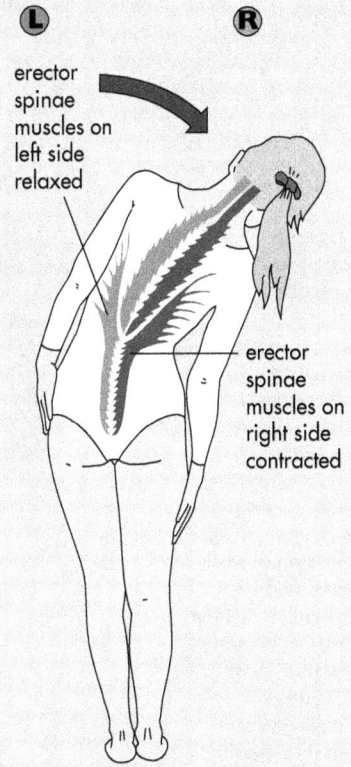

Ⓡ    Ⓛ    Ⓛ    Ⓡ

rectus abdominis muscles on left side relaxed

erector spinae muscles on left side relaxed

rectus abdominis muscles on right side contracted

erector spinae muscles on right side contracted

**Figure 1.16:** *Major muscles used in the lateral flexion, forward flexion and hyperextension of the trunk*

**Forward Flexion**

**Hyperextension**

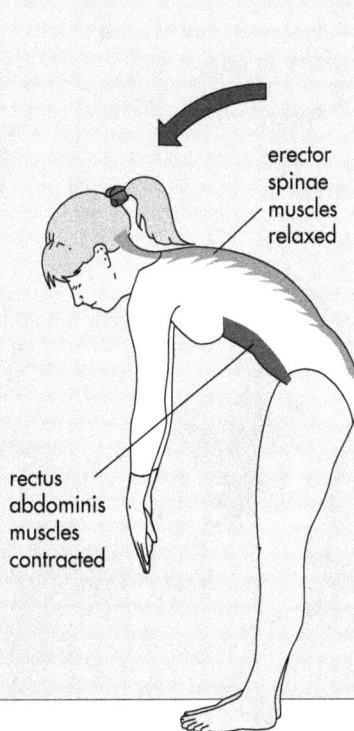

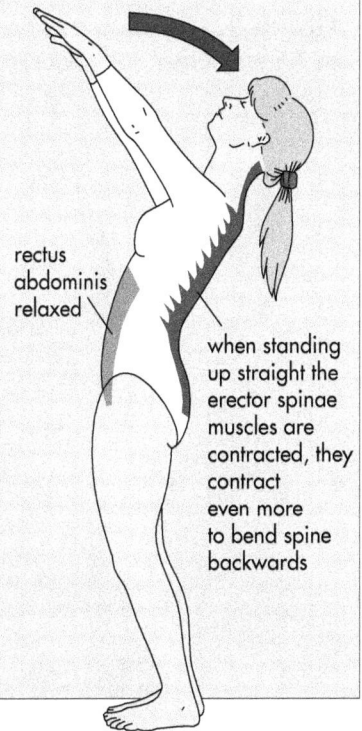

erector spinae muscles relaxed

rectus abdominis relaxed

rectus abdominis muscles contracted

when standing up straight the erector spinae muscles are contracted, they contract even more to bend spine backwards

**Major Muscles of the Upper Body**

Ⓛ **Back** Ⓡ   Ⓡ **Front** Ⓛ

trapezius attached to skull, vertebrae scapula, and clavicle; draws head back and to the sides, raises 'shoulders'

deltoids attached to scapula, vertebrae, and humerus; abducts and draws arm backwards and forwards

collar bone (clavicle)

Deltoids removed on left side

scapula

Trapezius removed on left side

left humerus

latissimus dorsi attached to vertebral column, ribs, & humerus; adducts arm, draws arm backwards and rotates it,

Pectoralis major removed on right side

Latissimus dorsi removed on right side

collar bone (clavicle)

Deltoids removed on left side

pectoralis major attached to sternum and humerus; adducts arm, draws arm forwards and rotates it

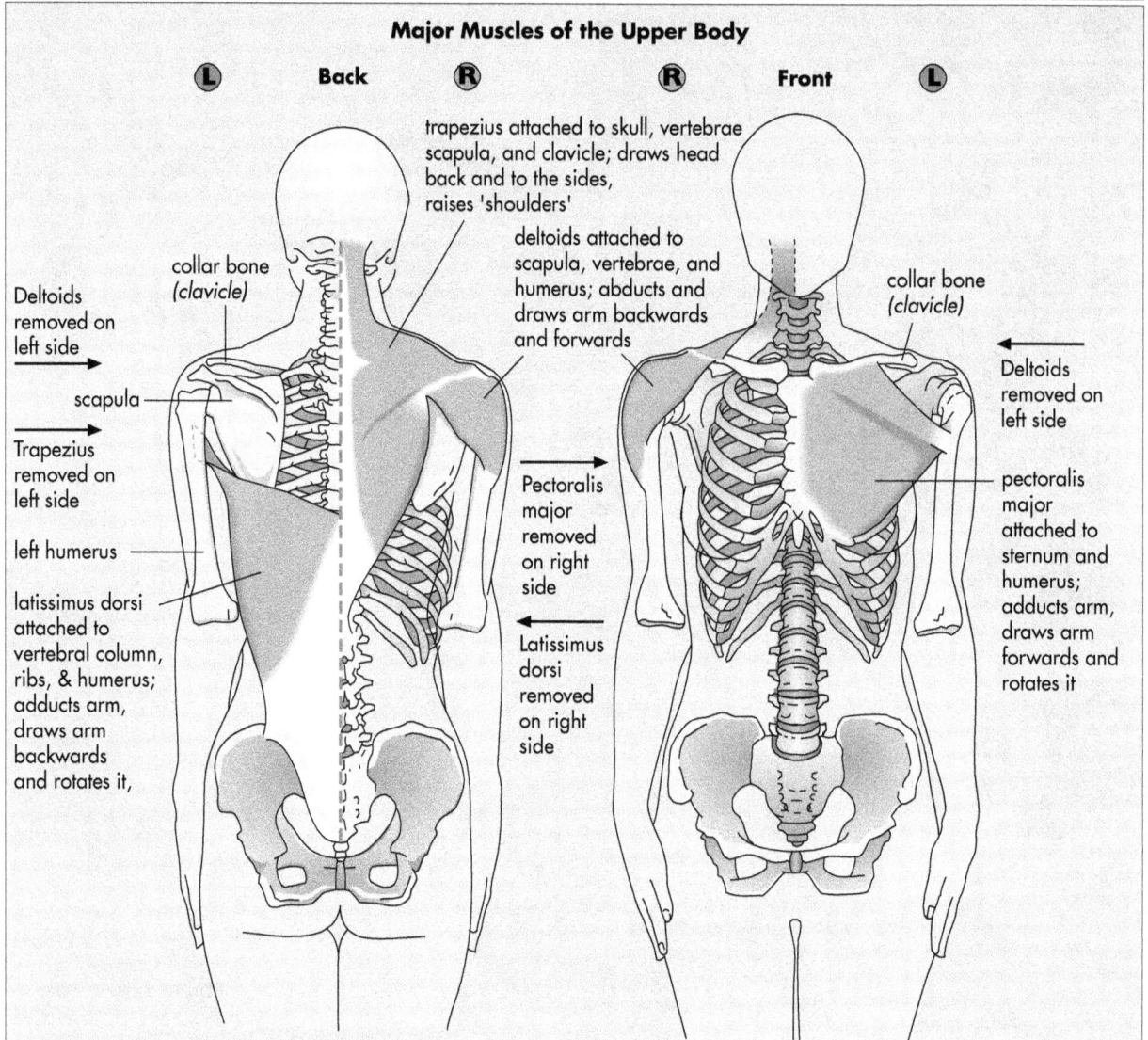

**Figure 1.17:** *Major muscles of the upper body used in the abduction and adduction of the upper arm.*

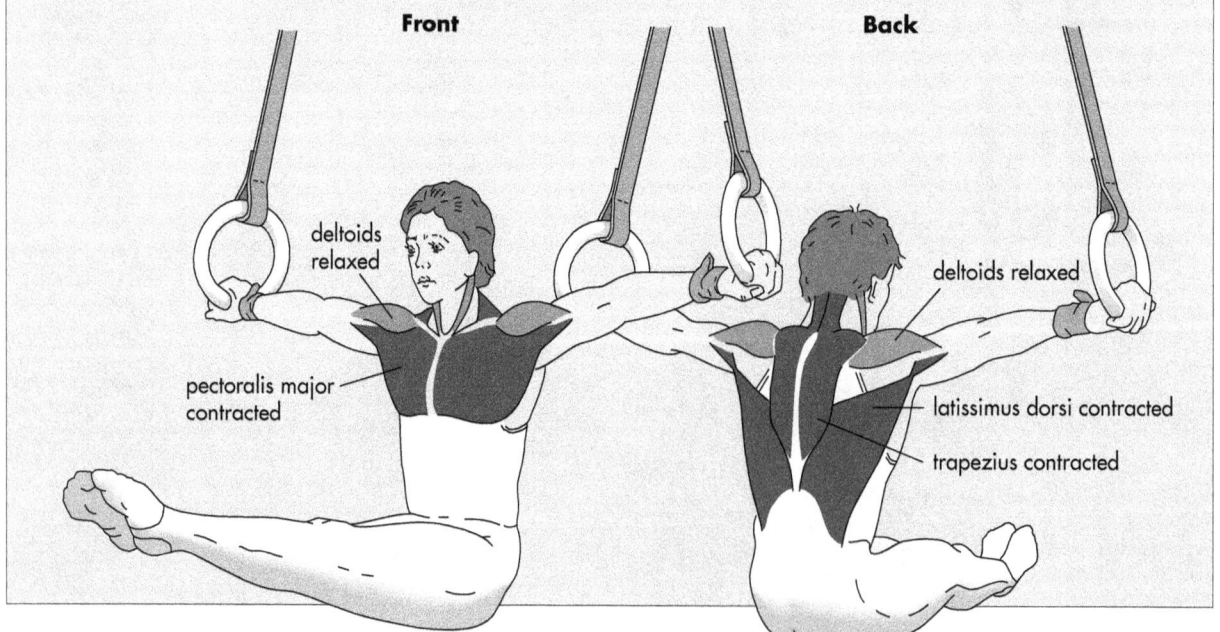

**Front**   **Back**

deltoids relaxed

pectoralis major contracted

deltoids relaxed

latissimus dorsi contracted

trapezius contracted

**Extension**
Right arm viewed from the side

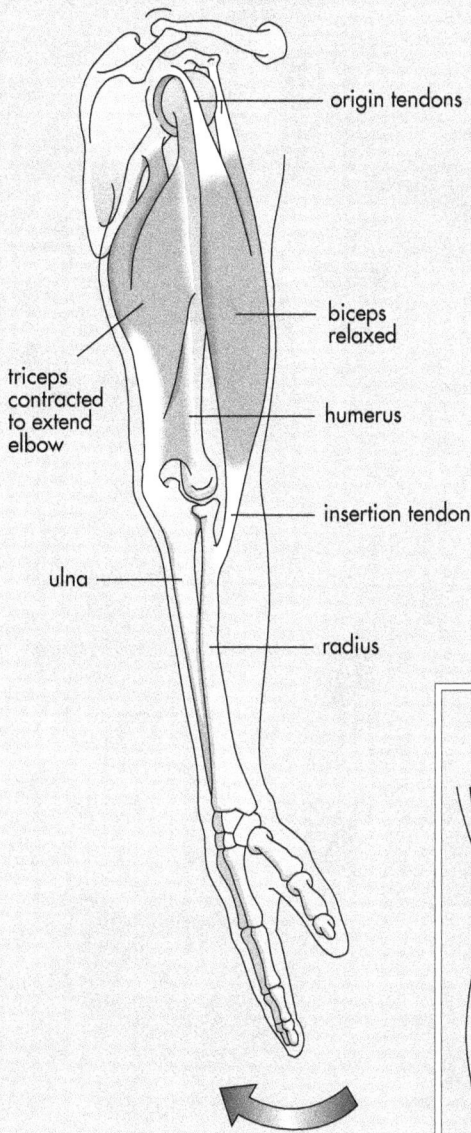

origin tendons

biceps
relaxed

triceps
contracted
to extend
elbow

humerus

insertion tendon

ulna

radius

**Flexion**
Right arm viewed from the side

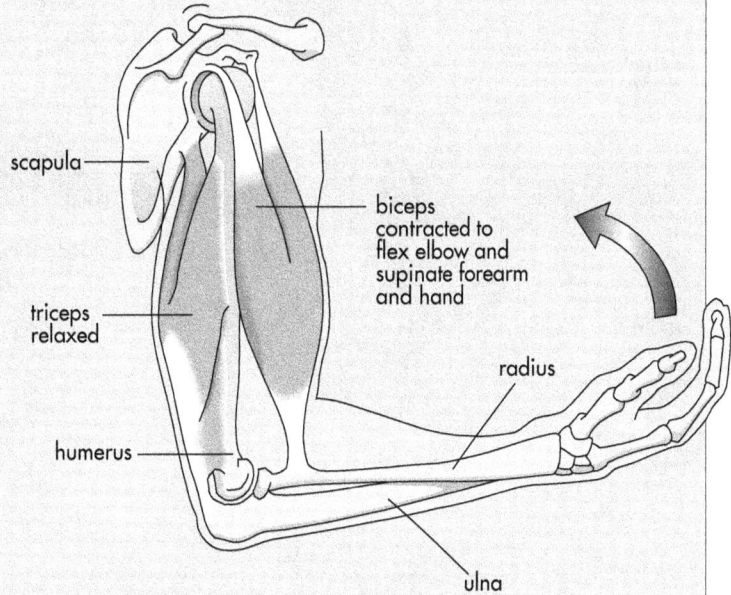

scapula

biceps
contracted to
flex elbow and
supinate forearm
and hand

triceps
relaxed

radius

humerus

ulna

*Figure 1.18: Major muscles used in the extension, flexion, supination and pronation of the arm*

**Supination (*Palm up*)**

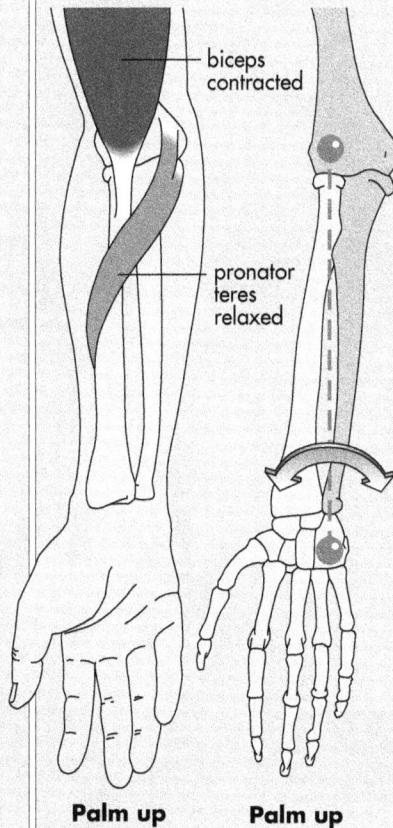

biceps
contracted

pronator
teres
relaxed

Palm up    Palm up

**Pronation (*Palm down*)**

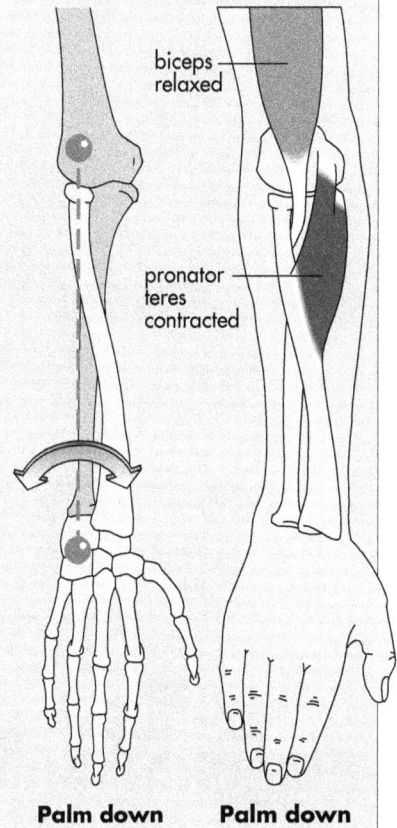

biceps
relaxed

pronator
teres
contracted

Palm down    Palm down

**Front View**

**Side View**

Ⓡ                    Ⓛ

tensor
fascia
lata

sartorius
attached to
pelvic girdle
and tibia;
contracts to
flex hip
and knee,
and rotate the leg

ilio-tibial band
of fascia
running from
pelvic girdle to
tibia

'knee cap'
(patella)

extensor digitorum
longus attached to
fibula and 'small toes'

gluteus medius

gluteus
maximus
attached to
pelvic girdle
and femur

'adductors'

'Quadriceps'
a group of
four muscles
attached to
pelvic girdle,
patella (knee cap),
and tibia;
contracts to
extend knee
and flex hip

'Hamstrings'

gastrocnemius
attached to femur
and heel bone

tibialis anterior
attached to tibia
and underside
of 'big toe'

soleus
attached to fibula
and heel bone

'Achilles' tendon

tensor
fascia
lata

fascia
lata

ilio-tibial band

'Quadriceps'
rectus femoris
vastus lateralis
in view

'knee cap'
(patella)

head of fibula
origin for
soleus and
extensor
digitorum
longus

shin bone
(tibia)

fibula

**Right leg seen from the right**

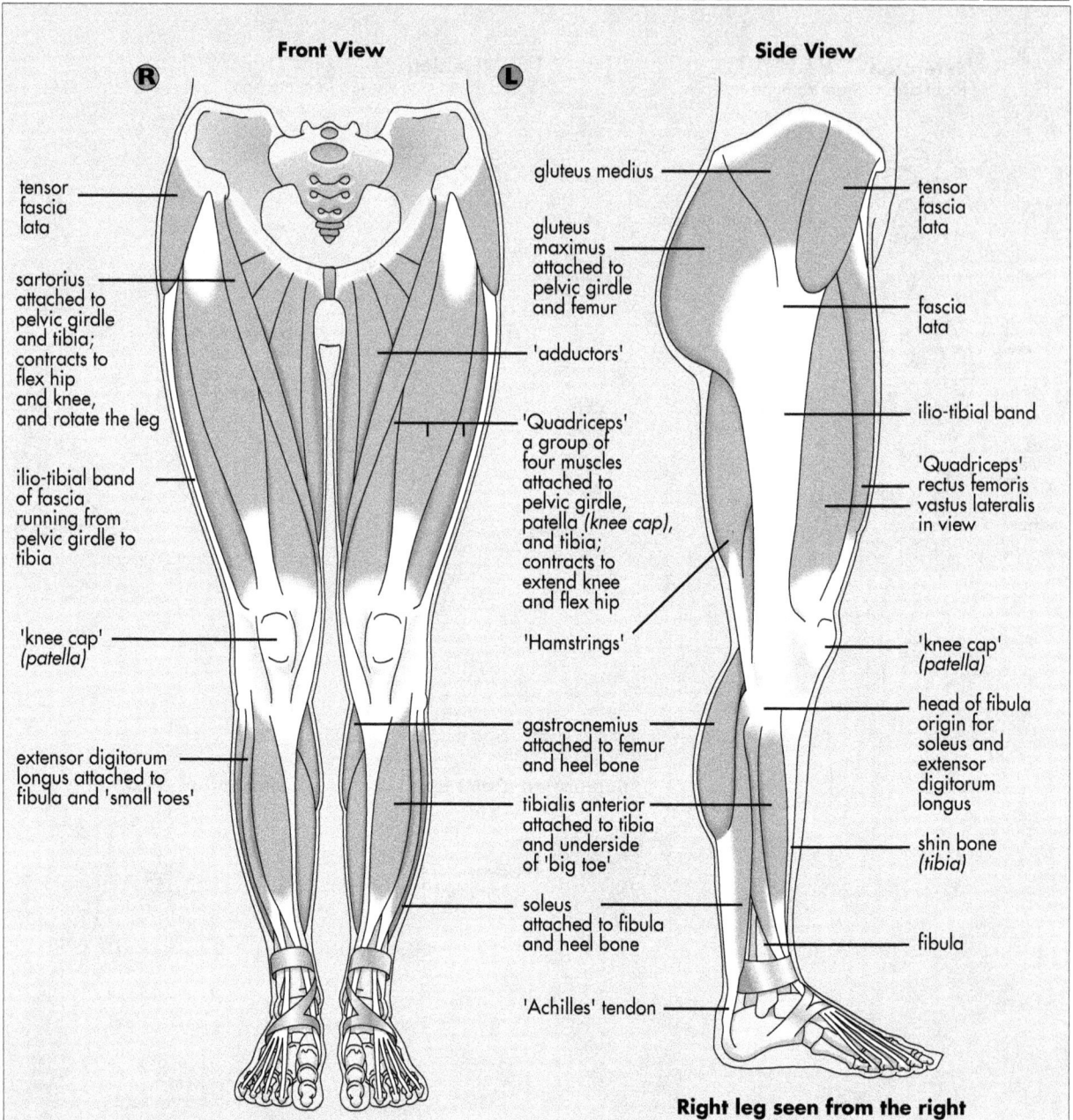

**Figure 1.19:** *Major muscles used in the leg
for extension and flexing of the hip and knee.*

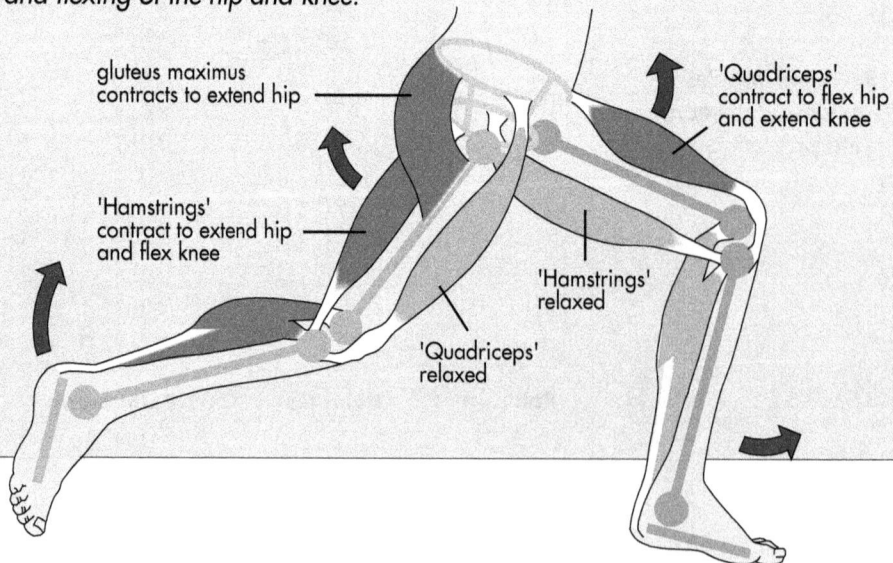

gluteus maximus
contracts to extend hip

'Hamstrings'
contract to extend hip
and flex knee

'Quadriceps'
contract to flex hip
and extend knee

'Hamstrings'
relaxed

'Quadriceps'
relaxed

**View from Back**

Ⓛ Ⓡ

gluteus maximus attached to pelvic girdle and femur *(covering gluteus medius)*

gluteus maximus removed to show gluteus medius attached to pelvic girdle and top of femur

hip joint

'Adductors' a group of muscles attached to pelvic girdle and upper femur

'Hamstring' muscles - group of three muscles attached to pelvic girdle, femur, tibia, and fibula - contract to extend hip and flex knee

gracilis attached to pelvic girdle and tibia contracts to adduct and rotate femur

femur

tibia

fibula

**View from Front**

Ⓡ Ⓛ

superficial adductors removed to reveal the adductor magnus

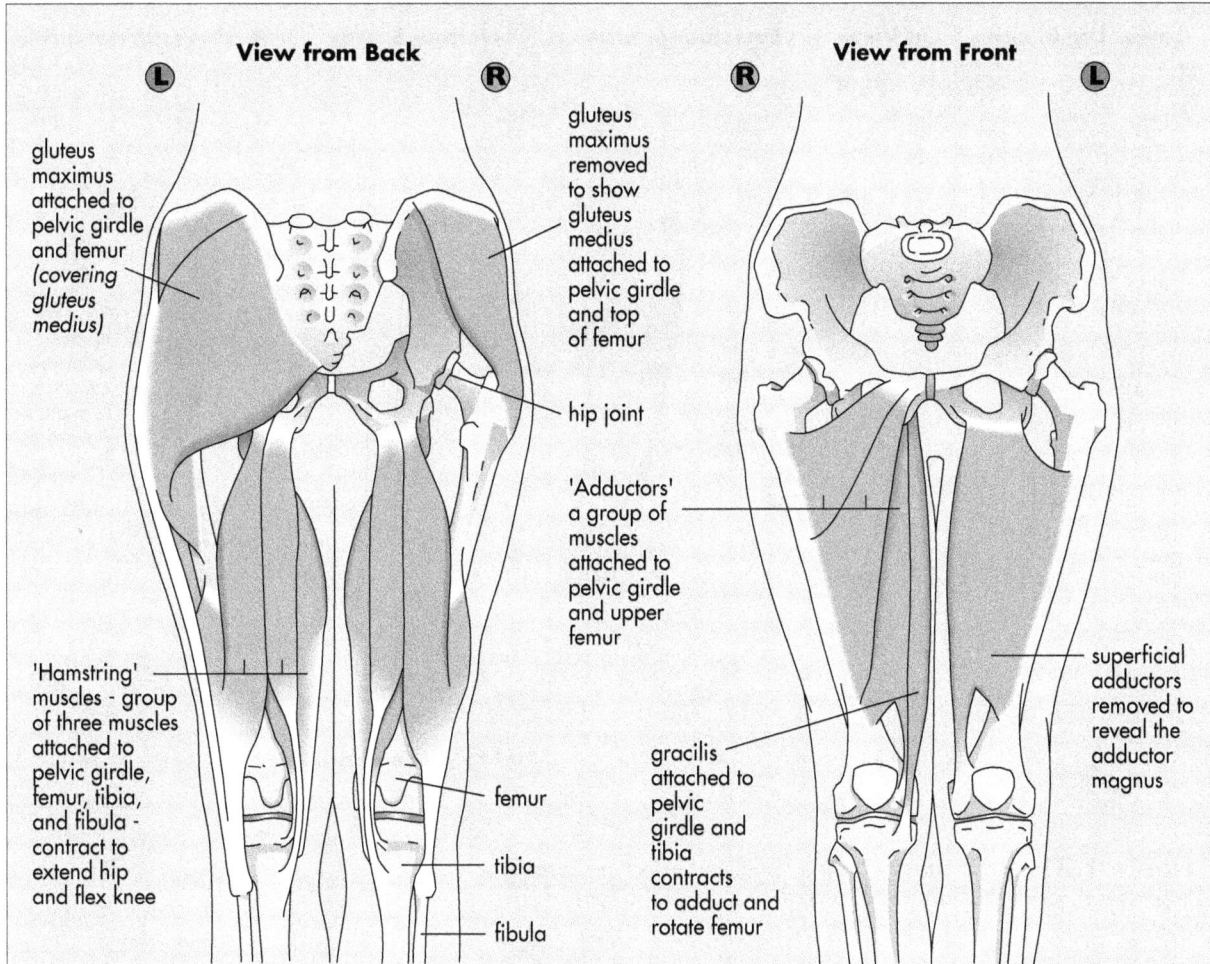

**Figure 1.20:** *Major muscles used, in abduction, adduction, flexion and rotation of the upper leg.*

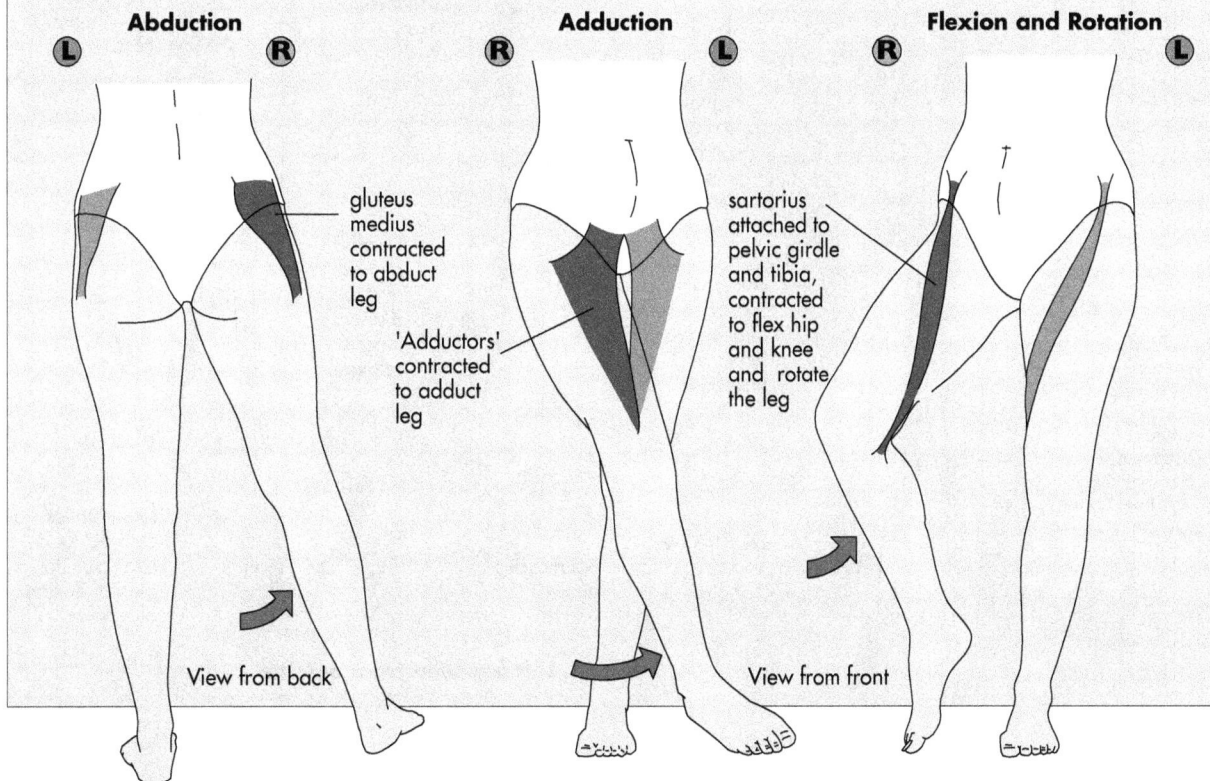

**Abduction**

Ⓛ Ⓡ

gluteus medius contracted to abduct leg

'Adductors' contracted to adduct leg

View from back

**Adduction**

Ⓡ Ⓛ

sartorius attached to pelvic girdle and tibia, contracted to flex hip and knee and rotate the leg

**Flexion and Rotation**

Ⓡ Ⓛ

View from front

**Lower Leg Muscles Front View**

Ⓡ　Ⓛ

**Eversion (pronation)**　　**Normal Stance**　　**Inversion (supination)**

extensor digitorum longus contracts to pronate (evert) foot

extensor digitorum longus — tibialis anterior

tibialis anterior contracts to supinate (invert) foot

tibialis anterior attached to tibia and underside of 'big toe'

extensor digitorum longus attached to fibula and 'small toes'

**Right leg seen from front**

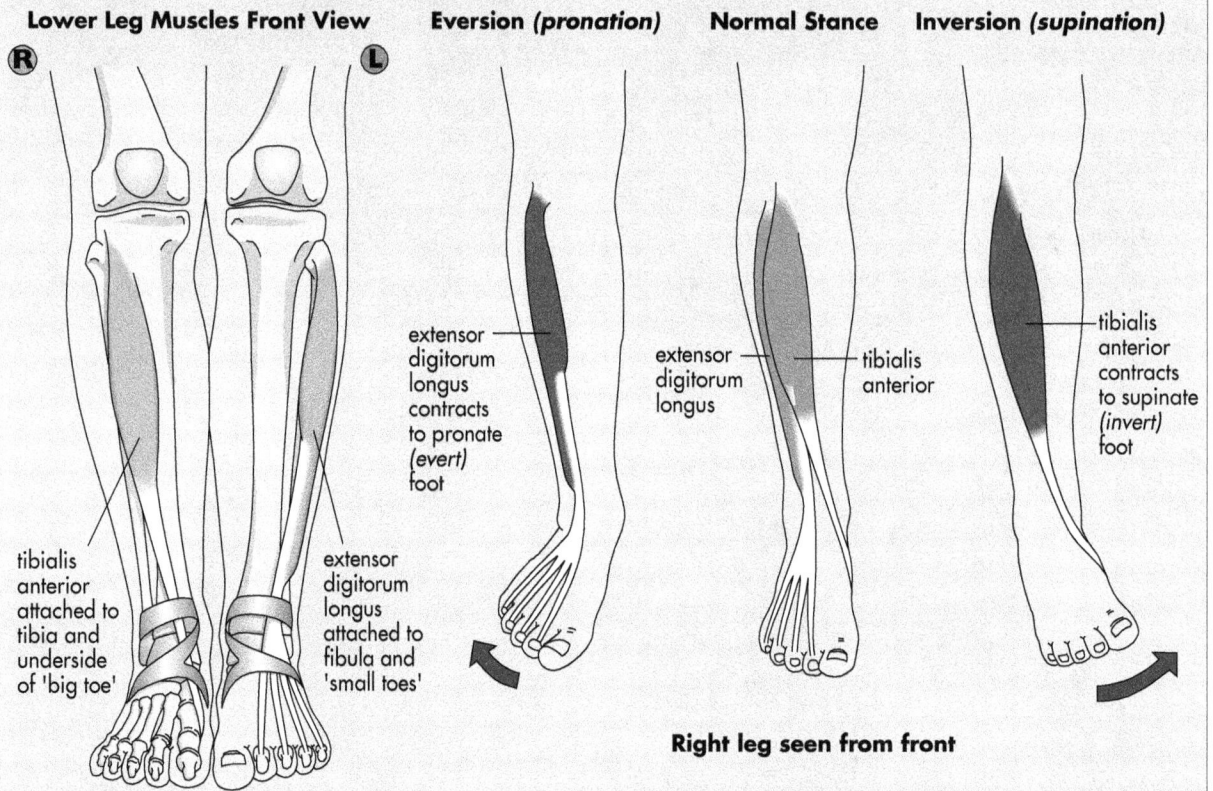

**Figure 1.21:** *Major muscles of the lower leg, used in pronation, supination, dorsi-flexion, and plantar-flexion of the foot.*

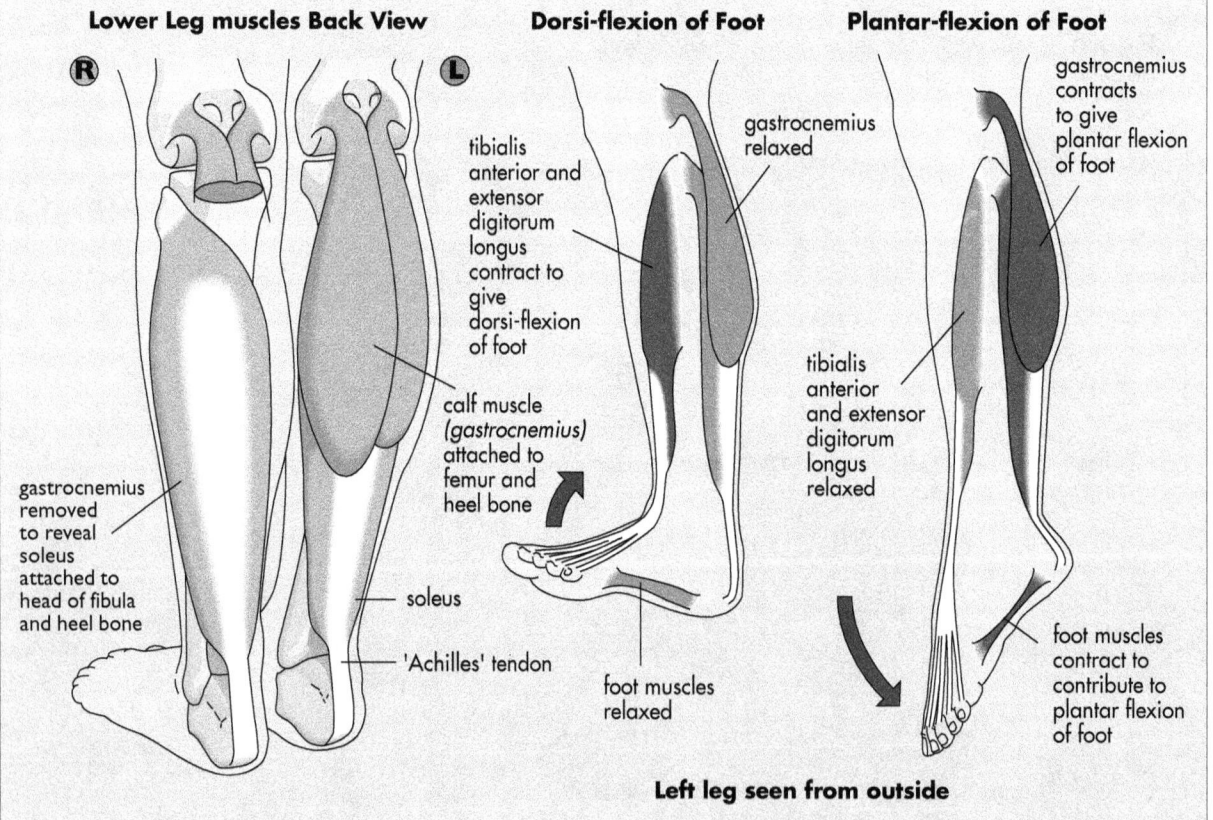

**Lower Leg muscles Back View**

Ⓡ　Ⓛ

**Dorsi-flexion of Foot**　　**Plantar-flexion of Foot**

tibialis anterior and extensor digitorum longus contract to give dorsi-flexion of foot

gastrocnemius relaxed

gastrocnemius contracts to give plantar flexion of foot

calf muscle (gastrocnemius) attached to femur and heel bone

tibialis anterior and extensor digitorum longus relaxed

gastrocnemius removed to reveal soleus attached to head of fibula and heel bone

soleus

'Achilles' tendon

foot muscles relaxed

foot muscles contract to contribute to plantar flexion of foot

**Left leg seen from outside**

**Table: 1.5:** *Major muscles and their actions*

| MUSCLE | ACTION | EXERCISE |
|---|---|---|
| Adductor longus | Adducts & rotates leg. | Forcing legs together. |
| Biceps brachii | Flexes elbow. Supinates forearm & hand. | Curling, chinning pull-ups. |
| Deltoids | Raise arm to shoulder level. Helps overhead movement. | Forward, lateral & back arm raises, overhead lifting. |
| Erector spinae | Holds spine erect *(extension)* Extends trunk backwards | Standing up straight, raising trunk while lying on front. Deadlifts |
| Gastrocnemius | Plantar flexion of foot. | Rising onto toes, as in running and jumping. |
| Gluteus maximus | Extends thigh & rotates it outwards. | Knee bending movements ie. squats, stair climbing, cycling. |
| 'Hamstrings' | A group of muscles which flex knee & extend hip. | Extending leg behind & flexing knee as in running. |
| Latissimus dorsi | Pulls arm downwards & backwards. Depresses shoulder girdle. | Pull downs behind head, rowing movements. |
| Pectoralis major | Pulls arm forwards & inwards. | All pressing movements. |
| Quadriceps femoris | A group of muscles which extends knee & flexes hip. | Knee bends, leg raises. |
| Rectus abdominis | Flexion of trunk. | Sit ups & leg raises. |
| Sartorius | Flexes knee, flexes hip & rotates thigh outwards. | Crossing legs whilst seated. |
| Soleus | Plantar flexion of foot. | Rising on tiptoe. |
| Sternocleido-mastoid | Pulls head forward & to side, rotates it. | Forward bridging, pushing on forehead. |
| Tibialis anterior | Dorsi-flexion & inversion of foot. | Running. |
| Trapezius | Draws head back & to sides, raises and lowers shoulders. | Shrugging, & all lifting overhead. |
| Triceps brachii | Extends elbow. | Press ups, dips, all overhead pressing. |

Major muscles in action.
Simple antagonistic actions become
complex when combined

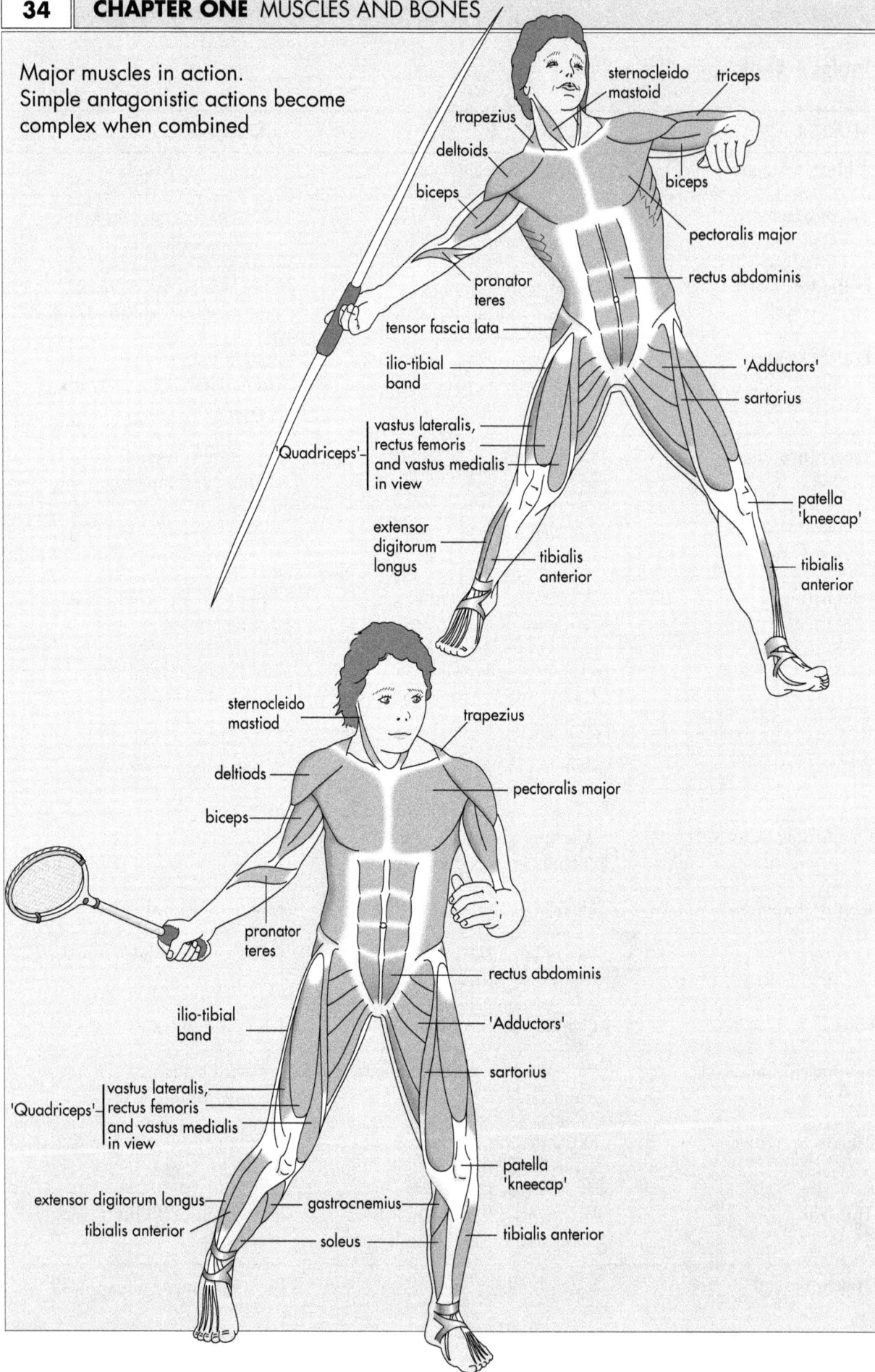

sternocleido mastoid

triceps

trapezius

deltoids

biceps

biceps

pectoralis major

rectus abdominis

pronator teres

tensor fascia lata

ilio-tibial band

'Adductors'

sartorius

'Quadriceps'

vastus lateralis,
rectus femoris
and vastus medialis
in view

patella 'kneecap'

extensor digitorum longus

tibialis anterior

tibialis anterior

sternocleido mastiod

trapezius

deltiods

pectoralis major

biceps

pronator teres

rectus abdominis

ilio-tibial band

'Adductors'

sartorius

'Quadriceps'

vastus lateralis,
rectus femoris
and vastus medialis
in view

patella 'kneecap'

extensor digitorum longus

gastrocnemius

tibialis anterior

tibialis anterior

soleus

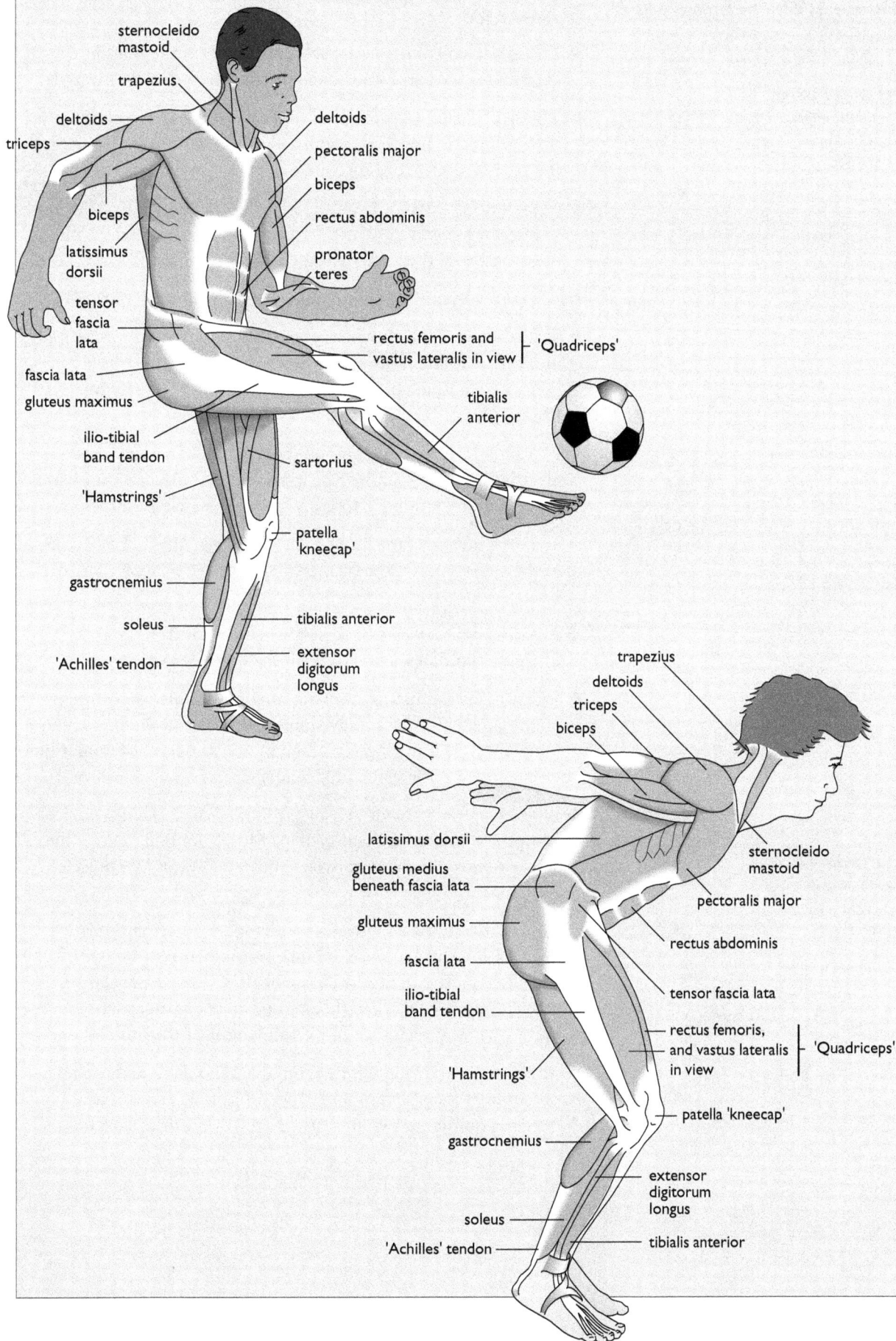

Figure 1 (upper figure labels):
- sternocleido mastoid
- trapezius
- deltoids
- triceps
- biceps
- latissimus dorsii
- tensor fascia lata
- fascia lata
- gluteus maximus
- ilio-tibial band tendon
- 'Hamstrings'
- gastrocnemius
- soleus
- 'Achilles' tendon
- deltoids
- pectoralis major
- biceps
- rectus abdominis
- pronator teres
- rectus femoris and vastus lateralis in view — 'Quadriceps'
- tibialis anterior
- sartorius
- patella 'kneecap'
- tibialis anterior
- extensor digitorum longus

Figure 2 (lower figure labels):
- trapezius
- deltoids
- triceps
- biceps
- latissimus dorsii
- gluteus medius beneath fascia lata
- gluteus maximus
- fascia lata
- ilio-tibial band tendon
- 'Hamstrings'
- gastrocnemius
- soleus
- 'Achilles' tendon
- sternocleido mastoid
- pectoralis major
- rectus abdominis
- tensor fascia lata
- rectus femoris, and vastus lateralis in view — 'Quadriceps'
- patella 'kneecap'
- extensor digitorum longus
- tibialis anterior

**Figure 1.23:** *The three orders of lever.*

**1st Order Lever**
Pivot (fulcrum) between effort and load
L – **P** – E

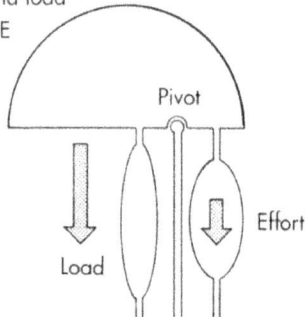

Pivot

Load

Effort

eg: atlas vertebrae to skull joint

**2nd Order Lever**
Load between effort and pivot
E – **L** – P                          eg: ankle joint

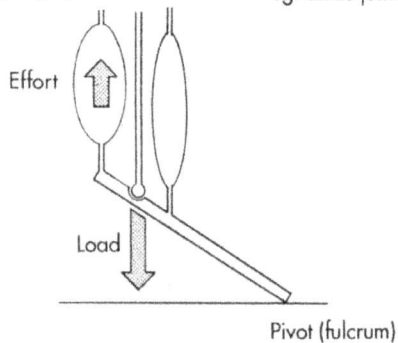

Effort

Load

Pivot (fulcrum)

**3rd Order Lever**
Effort between load and pivot
P – **E** – L                          eg: elbow joint

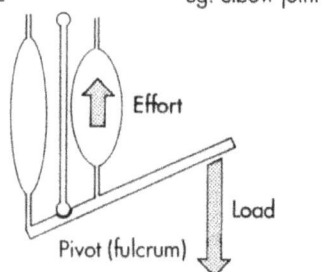

Effort

Load

Pivot (fulcrum)

| Order of lever | 1 | 2 | 3 |
|---|---|---|---|
| Middle initial | P | L | E |

## SUMMARY

❖ The arrangement of the muscle fibres within the muscle determines the force and range of contraction of that muscle.

❖ The structural and functional unit of a skeletal muscle fibre is the sarcomere, which is composed of overlapping filaments of the proteins actin and myosin.

❖ Cross bridges arising from the myosin filaments move the actin filaments one way towards the centre of the sarcomere, thus drawing the Z discs closer together, shortening the sarcomere.

❖ ATP provides all the energy necessary for muscle contraction.

❖ Slow twitch *(Type I)* muscle fibres have a greater blood supply, more myoglobin, and more mitochondria; giving them a greater oxidative potential than Fast twitch *(Type II)* muscle fibres.

❖ Fast twitch fibres possess more contractile filaments, more glycolytic enzymes, more phosphocreatine, and are initiated more rapidly and more forcibly than slow twitch fibres.

❖ Training increases the size of muscle fibres, affects both types in the same way and may alter their relative proportions, with time.

❖ A motor unit consists of a motor neurone and the muscle fibres it stimulates, which are all of the same type.

❖ Fatigue may be due to events related to stimulation of the muscle, depletion of energy supplies, effects on the brain, fluid imbalance, overheating, shortage of oxygen and accumulation of lactate, and the associated acidic effects.

❖ Proprioceptors provide sensory feedback or 'kinaesthenic feel' from the musculo-skeletal system. Muscle spindles detect the state of tension within muscles, and are used in fine muscle control.

❖ Concentric action of muscles results in a shortening of the muscle, and can be classified into isotonic and isokinetic. Eccentric muscle action occurs when the muscle develops its active tension whilst lengthening. Isometric muscle action occurs without a change in overall length of the muscle.

❖ Muscles act in groups, for example antagonistic pairs. The outcome of the strength and range of muscle contraction depends on the particular muscle's relation to the skeleton.

# 2 ENERGY RELATIONS IN ACTION

## OBJECTIVES

To enable the reader to understand the following.

❖ The sources of energy available for contracting muscle.

❖ The role of hormones in the use of energy.

❖ The particular importance of energy rich phosphate compounds in muscle contraction, their use and regeneration.

❖ The use of fats, carbohydrates and proteins as energy sources during exercise.

❖ The formation of lactic acid under conditions of shortage of oxygen supply, its accumulation and subsequent removal.

❖ The concept of energy efficiency.

❖ Heat balance and fluid loss during exercise.

## 2.1 SOURCES OF ENERGY

The energy rich materials in the diet, namely fats, carbohydrates, and proteins, are the source of all energy within the body.

Apart from the need for sufficient carbohydrates, and the special conditions relating to carbohydrate loading, there are no agreed views on how special diets may improve performance. A balanced diet will sustain virtually any activity.

The end products of digestion, namely glycerol and fatty acids from fats, monosaccharides e.g. glucose from complex carbohydrates such as starch, and amino acids from proteins, are absorbed into the blood stream, and undergo a range of different processes, according to the demands of the body.

### 2.1.1 GLYCEROL AND FREE FATTY ACIDS

The glycerol and free fatty acids *(FFAs)* may be used directly as a source of energy, or resynthesised into fats and stored in the fatty *(adipose)* tissue which contain as much as 98% of the body's total energy reserves. Glycerol *(but not FFAs)* can be converted into glucose. This is particularly important when the diet is low in carbohydrates, and during endurance exercise which depletes the glycogen reserves.

### 2.1.2 GLUCOSE

Some of the glucose may be used directly from the blood plasma, and any glucose in excess of the normal levels in the blood is converted either into the insoluble storage carbohydrate glycogen in the muscles and the liver, or into fat which is stored in specialised adipose tissue under the skin, and around the major organs of the body.

### 2.1.3 AMINO ACIDS

The amino acids also circulate in the plasma, and are mainly used to synthesise proteins for growth and repair throughout the body. Excess amino acids cannot be stored in the body, so they are deaminated in the liver, a process in which their nitrogen containing amino groups are removed, producing toxic nitrogen containing waste products such as urea. Whilst exercising, the urea excreted in the sweat exceeds that excreted by the kidneys, which have a reduced blood supply during exercise. The remainders of the now deaminated amino acid molecules, can either be used directly in the respiratory pathways, or be converted into glucose.

### 2.1.4 INTERCONVERSIONS OF ENERGY SOURCES

All these fuels are to a certain extent interconvertible, depending on their relative supplies, and the particular demands of the body at any one time. The liver is the main centre for these interconversions, and for maintaining constant optimum *(homeostatic)* levels in the blood. Theoretically most of these interconversions are reversible, but this is not always true in practice. Thus although the liver can convert excess glucose more readily into fats than into glycogen, very little glucose, if any, can be produced from fats. Also, although liver glycogen is readily converted into blood glucose, muscle glycogen is not.

## 2.2 USE OF ENERGY SOURCES IN EXERCISE

At rest, about two thirds of the energy required for the body is derived from fats, and one third from the carbohydrates glycogen and glucose. The relative amounts of these fuels used in exercise depends on many factors; including the type of exercise *(whether it is intermittent or continuous, short or long, light or heavy)*, the types of muscle fibre involved, the nature of the diet, and the state of physical training - which in turn determines the supply of oxygen.

### 2.2.1 FATS

The complete oxidation of fats depends on the presence of oxygen, so the relative contribution of fat to the energy demands of the body is largely determined by the supply of oxygen to the tissues. In the long term functioning of slow twitch muscle fibres, fat is the main energy source under aerobic conditions, so that after about an hour of continuous aerobic exercise the majority of the energy is derived from fats. Training increases the capacity for oxygen uptake, and therefore increases the ability to use fats.

Very little fat is stored in the muscle fibres, therefore the fatty acids and glycerol must be supplied via the bloodstream. The adipose tissue releases free fatty acids *(FFAs)* into the plasma of the blood, which are transported to the muscles loosely bound to plasma proteins. Those who take part in regular aerobic endurance work increase their ability to release fatty acids from their fat stores, and can therefore use more fats as an energy source during exercise, thus conserving or 'sparing' glycogen reserves.

The glycerol from the breakdown of fats can either enter the carbohydrate respiratory pathways directly, or be converted into glucose which is then broken down in the normal way.

Fats tightly bound to proteins, known as lipoproteins are also transported in the plasma. These originate from dietary fats absorbed in the small intestine. They are carried to the liver where they are converted to two different types of lipoprotein, low density lipoproteins *(LDLs)* which supply cholesterol to the tissues, and high density lipoproteins *(HDLs)* which remove cholesterol from the tissues. Exercise tends to increase the ratio of HDLs to LDLs , which seems to be a factor in lowering the risk of atherosclerosis *(the formation of fatty deposits or plaques on the walls of blood vessels)*. Their exact role in the energy economy of the tissues is not fully understood, but the enzyme lipoprotein lipase, which is attached to the surface of the cells lining the capillaries supplying the muscles, releases FFAs from these circulating lipoproteins, making them available to the working muscles.

### 2.2.2 CARBOHYDRATES

Glycogen within the muscle fibres is available for immediate use, for example at the onset of exercise and/or during intense exercise. During these periods, the supply of oxygen is usually insufficient for the complete aerobic oxidation of carbohydrates and fats. Only carbohydrates can be broken down anaerobically to maintain the supply of energy to the contracting muscle fibres.

---

### ❖ *Training and Performance Applications*

*Generally speaking, as the intensity of the exercise increases there is an increase in the use of carbohydrates as an energy source, so that they become the main fuel at about 50 - 70% of the maximum oxygen uptake ( $\dot{V}_{O_2}$ max). Elite marathon runners can perform at more than 80% of their maximum oxygen uptake and therefore use mainly glycogen, even though they are performing what would normally be considered as endurance exercise mainly exploiting fat.*

---

The anaerobic breakdown of carbohydrates *(anaerobic glycolysis)* results in the production of lactic acid *(Section 2.5.2, p 48)*. The increased acidity *(decrease in pH)* inhibits the mobilisation of FFAs and subsequently interferes with their use, further emphasising the importance of the muscle glycogen stores as an energy source during intense activity.

When oxygen supplies are sufficient, the use of carbohydrates as a fuel has another advantage, that is, for a given volume of oxygen, carbohydrates yield about 12% more energy than fat.

Carbohydrate Stores
The total carbohydrate store in the body is about 375-475 g; of which up to 345 g are stored as muscle glycogen, 90-110 g are stored as liver glycogen, and 15-20 g are found in the plasma as blood glucose. Active muscle fibres can use glucose obtained from all these sources. However, blood glucose is the only source of energy for the brain, and most of the glucose released from the liver glycogen stores is used for this purpose.

The main disadvantage of using glycogen is that stores are very limited when compared with fat, and they can be depleted relatively quickly within a few hours. The depletion of muscle glycogen stores leads rapidly to muscular fatigue. This is why the use of fat is of such importance the longer exercise continues.

During the recovery period there is a rapid replenishment of glycogen within the first few hours, which is quicker in the fast twitch than in the slow twitch muscle fibres. After this, it can take two or more days for it to be restored fully to its pre-exercise levels.

### ❖ *Training and Performance Applications*

*Glycogen stores can be increased to higher levels than normal by special dietary and exercise regimes. This process of glycogen 'loading' or 'supercompensation' can help extend the period of endurance exercise before the onset of fatigue. The technique is based on the principle that if glycogen stores are depleted, and held at a low level for a few days, either by restricting the intake of carbohydrates and/or continuing to exercise, then when carbohydrates are readily available once again in the diet, the body overcompensates and increases its stores of glycogen above the normal level. However, problems can arise with the technique. It is not always successful, declines in effectiveness with repeated use, and may even disrupt normal performance. The rapid increase in weight associated with glycogen loading, is due to the volume of water that is associated with the stored glycogen.*

### Maintenance of Blood Glucose Levels

If the blood glucose levels begin to fall, brain function is interfered with. Therefore there are safeguards that prevent the muscles from using blood glucose too readily. For example, the ability of the muscles to absorb glucose from the blood is decreased by the reduction in the level of insulin in the blood, which occurs if the blood glucose falls. Therefore, if blood glucose begins to fall, the muscle fibres are decreasingly able to take up glucose and use it as an energy source. Therefore muscles will use glucose from their own glycogen stores, rather than from the plasma.

### ❖ *Training and Performance Applications*

*During prolonged exercise, as muscle glycogen is depleted, exercising muscles do increasingly use blood glucose. A variety of homeostatic or steady state mechanisms, most of which are centred on the liver, operate to maintain blood glucose at optimum levels for as long as possible. For example as exercise proceeds, breakdown of liver glycogen releases glucose into the blood plasma. As the liver glycogen is depleted, glucose is increasingly made from new sources by the liver (gluconeogenesis) stimulated by the hormone glucagon. This 'new glucose' can be produced from glycerol (from fats), and/or from amino acids derived from proteins (particularly alanine, which may come from the breakdown of muscle protein itself). Eventually, if the use of glucose by the muscles and the brain becomes greater than the supply from the liver, the blood glucose level will gradually drop (hypoglycaemia), causing fatigue and a decrease in the normal functions of the brain. This is illustrated by the fact that athletes can have great difficulty in making the simplest of calculations in the last few miles of a Marathon.*

*During exercise, the taking in of glucose does not necessarily help. Concentrated solutions remain in the stomach and do not pass readily into the small intestine, where most absorption of water and glucose into the blood takes place under these conditions. Also, if the solution is too concentrated, even if it does pass into the small intestine absorption of water is reduced and in fact water will tend to be drawn out of the blood by osmosis. However, the absorption of water into the blood is increased if the solution of glucose is the same concentration as the blood (isotonic), this in fact is the main role of such isotonic drinks, rather than energy replacement. During endurance exercise, the problems associated with the intake of energy and water may be overcome by the use of a solution of long chain carbohydrate molecules (maltodextrins), obtained from the partial breakdown of starch. This does not slow the rate of stomach emptying, nor does it oppose the uptake of water in the small intestine, or trigger the release of insulin (Section 2.3), and it provides energy.*

### 2.2.3 PROTEINS

It is estimated that up to about 10% of the energy used in exercise may be derived from proteins. There are several pathways by which this can occur.

Amino acids excess to requirements for protein synthesis can have their amino groups removed *(deamination)* in the liver, these amino groups produce urea, and the remainders of the molecules are broken down in respiration with the release of energy, or resynthesised into glucose and/or glycogen.

As the intensity of exercise increases, protein in the muscles is broken down into amino acids, and the amino acid alanine is increasingly released from the exercising muscles into the blood. The alanine is transported in the blood plasma to the liver, where it is deaminated *(producing urea)* and the residue is converted to glucose, which can then return to the exercising muscles, or be used to replenish liver glycogen stores. In this way muscle protein can be used as an energy source.

## 2.3 HORMONAL REGULATION OF ENERGY METABOLISM

Hormones are 'chemical messengers' secreted by ductless *(endocrine)* glands directly into the blood stream. They control or influence all aspects of metabolism. With respect to energy metabolism five hormones are mainly involved.

Adrenaline, **noradrenaline** and **glucagon** stimulate the conversion of glycogen to glucose, and together with **growth hormone** they stimulate the conversion of fats to FFAs and glycerol. These hormones also increase the activity of the enzyme lipoprotein lipase which increases the rate of supply of fatty acids. An increase in the secretion of these hormones will therefore increase the availability of energy for exercise.

**Insulin** inhibits the effects of adrenaline, and glucagon, and also inhibits the enzyme lipase which catalyses the breakdown of fats into FFAs and glycerol. An increase in the secretion of insulin will therefore decrease the availability of energy for exercise. An anabolic or protein building effect is also claimed for insulin, and some power athletes attempt to maintain raised insulin levels by continuous carbohydrate 'snacking' during workouts.

---

❖ *Training and Performance Applications*

*It has been suggested that consuming sugars within an hour of starting exercise could be counterproductive, as the secretion of insulin would be stimulated, and the available energy for exercise therefore decreased. However, in practice this does not appear to be the case. Once the activity has started, the likelihood of any insulin reaction that might occur decreases, as a result of noradrenaline supressing insulin activity; and there is certainly an advantage to be gained by taking a certain amount of 'sugars' during the activity not only as a source of energy but also to aid the uptake of water by the gut.*

---

## 2.4 ENERGY SUPPLY SYSTEMS

Broadly, there are three systems involved in supplying energy to the working muscles. The **ATP/PC** system or phosphagen system, the **lactic anaerobic** system or anaerobic glycolytic system, and the **aerobic** system. Each of these is suited to a particular type of exercise, and each can be improved by specific training regimes. However, there is also extensive overlap between the systems which share common metabolic pathways.

The end product of all the energy releasing processes of cellular respiration is the energy rich adenosine triphosphate or ATP. ATP is the energy currency of all living things. It is the form in which energy is supplied for all purposes. Each molecule contains three phosphate groups, the end one of which can be more easily displaced than the others, with energy being made available for use by the muscle fibres. In this process, adenosine diphosphate (ADP) and a free phosphate group (P) are formed. This process is

**Figure 2.1:** *Energy flow through ATP and PC (Phosphagen system).*

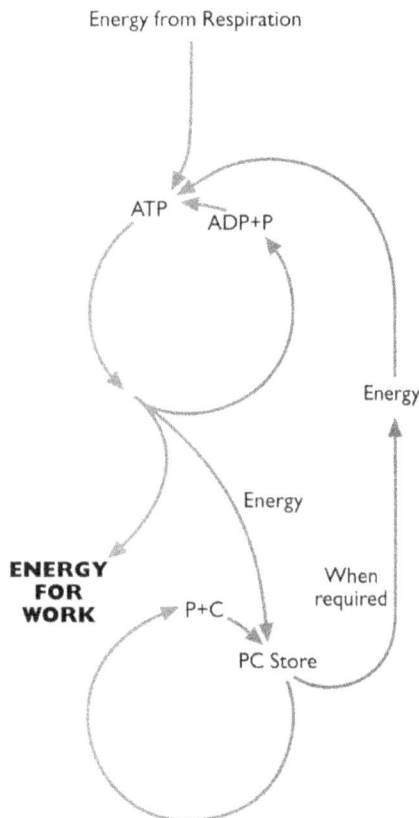

reversible, and energy released from the oxidation of energy rich materials can be trapped in the ATP molecule when a phosphate group is combined with ADP to reform ATP.

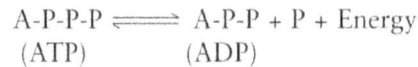

$$\text{A-P-P-P} \rightleftharpoons \text{A-P-P} + \text{P} + \text{Energy}$$
$$(\text{ATP}) \qquad (\text{ADP})$$

The relative amounts of ATP and ADP + P control the rate of breakdown of energy rich substances in respiration. If ADP and P are present, then energy is required to synthesise ATP, and the rate of respiration increases to supply this energy. On the other hand, if there is only ATP present, there is no immediate demand for energy, and the rate of respiration decreases.

### 2.4.1 SOURCES OF ENERGY FOR THE REGENERATION OF ATP

There is a only a limited store of ATP within the muscle fibres, and as all energy used by the body must be channelled through ATP, it must be replenished as soon as stores become low. To ensure its rapid regeneration there is another energy rich phosphate compound in the muscles, namely phosphocreatine *(PC)*, which acts as the most immediate energy reserve for the resynthesis of ATP. The rapid availability is particularly important in fuelling muscular work at the onset of exercise, especially contractions of high power.

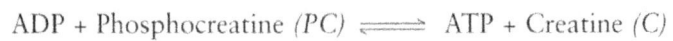

$$\text{ADP} + \text{Phosphocreatine } (PC) \rightleftharpoons \text{ATP} + \text{Creatine } (C)$$

PC in turn can only be regenerated using energy from ATP. Therefore, during maximal exercise, PC cannot be replenished, as the ATP is being used continuously in muscle contraction. However, it can be replenished within a few minutes after such exercise stops, when ATP becomes available again after having been regenerated via mainly aerobic pathways.

Energy from Respiration

ATP        ADP+P

Energy

Energy

**ENERGY FOR WORK**

When required

P+C

PC Store

---

❖ *Training and Performance Applications*

*The ATP/PC or phosphagen system provides enough energy at a sufficient rate for up to 6-10 seconds of flat out effort, as for example in the 100 metres sprint with most of the ATP store being depleted within the first few seconds. After such effort the PC stores are depleting, and represent the limiting factor for muscle contraction during short term exercise. The phosphagen system is also of major importance in games playing, which involves repeated short bursts of intense activity with periods of recovery in between. Training has no effect on the total amount of ATP available, but the amount of PC can be increased by speed training. This has the acute effect of delaying the onset of the use of the anaerobic glycolytic system to regenerate the ATP. This means that the reduction in power output associated with the shift from the rapidly available energy of stored ATP/PC, to the indirect (and potentially more fatiguing) regeneration from anaerobic glycolytic pathways, is delayed. The formation of PC is limited by the availability of creatine, which enters the body from foods such as meat*

Levels of ATP and PC
during a flat out sprint.

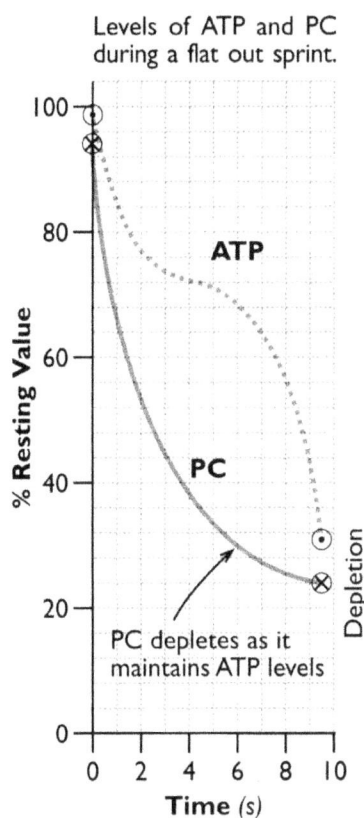

❖ **Training and Performance Applications**

*All three components of the energy supply system, the ATP/PC, the anaerobic, glycolytic, and the aerobic, operate from the outset of any event involving continuous movement, but to varying degrees. For example, in a hard run 800 metres race, the ATP/PC system will fuel the quick kick off from the start. The anaerobic glycolytic system will be mainly used in the first 200-400 metres, which is the fastest phase of the race when demand for oxygen is greater than the supply. The runners gather themselves for the final lap, and by now the respiratory and circulatory systems have adjusted to the demands, so that aerobic respiration can increase. In the final kick for home, demand for oxygen again exceeds supply and the anaerobic glycolytic system floods the body with lactic acid, bringing the athlete to a virtual standstill at the end of the race, hopefully just past the finish line.*

*and fish. Creatine content of such foods will be sufficient for most activities, however, methods have been employed to extend the bodily storage facility, for which major claims for training and performance have been made. Creatine supplementation has been suggested with doses of 20-30g per day, consumed in 4-5 equal doses, for 5-7 days. This is known as the loading phase and will increase the total muscle creatine content. There is, however, an upper limit to intramuscular creatine content past which excess is excreted by the kidneys. This is why the loading phase is followed by a maintenance phase, with a dose of only 2-3g per day in order to preserve creatine stores. The increase in total muscle creatine of about 20% results in a significant increase in the amount of work which can be performed close to the maximum. It also results in a quicker recovery rate after exercise.*

Although PC is always used to some extent at the beginning of any type of exercise to regenerate ATP, ATP is regenerated from other sources as well. If the exercise is of the speed endurance type, energy from anaerobic glycolytic respiration *(in which lactic acid is generated)* is mainly used to regenerate the ATP.

If the exercise is of the endurance type, where the oxygen delivery systems become capable of sustaining aerobic respiration *(as a result of an increase in the breathing and heart rates)*, more energy from the aerobic oxidation of glycogen, FFA and blood glucose is used as the slow twitch muscle fibres become more involved. With this system a steady state can be reached, in which ATP is regenerated at the same rate as it is broken down, and performance can be maintained for long periods.

During exercise some ATP can also be regenerated from the combination of ADP molecules:

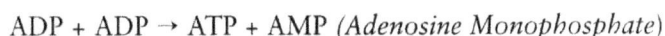

$$ADP + ADP \rightarrow ATP + AMP \text{ (Adenosine Monophosphate)}$$

AMP activates respiratory enzymes, and stimulates the secretion of adrenaline, both of which increase the rate of respiration and therefore the regeneration of ATP from aerobic sources.

It is important to note that all these ways of resynthesising ATP may operate at the same time, but the relative contribution of each one depends upon the activity and its duration.

## 2.5 METABOLIC PATHWAYS

The energy that is trapped and stored in the phosphagen system originates from the energy rich respiratory substrates. It is released from the energy rich sources by a series of enzyme catalysed reactions. In these metabolic pathways the respiratory substrates are oxidised to carbon dioxide and water, in the process known as cellular respiration. The breakdown of the different respiratory substrates involves common pathways. The main pathway is that of the breakdown of carbohydrates, with the breakdown of fats and proteins linking in at different points.

❖ *Training and Performance*
   *Applications*

*The complete oxidation of fats depends on carbohydrate breakdown, which supplies intermediate molecules for the complete breakdown of fats. This is the origin of the phrase 'fats only burn in the flames of carbohydrates'. When not enough carbohydrate breakdown occurrs, eg when the glycogen reserves are depleted, the incomplete breakdown of fats can result in toxic ketosis, which may cause breathlessness and vomiting. The depletion of glycogen reserves explains the experience of 'hitting the wall' in endurance events such as the Marathon (26.2 miles, 41.9 km), where it is said that when you reach 20 miles (32k) you are only half way. Physiologically and psychologically this statement has an element of truth! The intake of **caffeine** increases the blood levels of FFAs and the ability of the muscles to use them, and thus spares glycogen. Caffeine stimulates the nervous system increasing alertness and reducing pain and fatigue. It also increases the heart rate and ventilation rate. All these affect sport activity. Although found in modern diets (coffee, tea, cocoa, chocolate, and colas) caffeine above a certain level is considered as doping, where doping is defined as "the administration or use of substances in any form alien to the body, or of physiological substances in abnormal amounts and/or by abnormal methods by healthy persons with the exclusive aim of attaining an artificial and unfair increase in performance in sports." Caffeine is banned at a level equivalent to 8 strong cups of percolated coffee or 10 cups of strong tea in two hours. Adverse effects include increased water loss in urine, anxiety, insomnia, nausea, diarrhoea, headaches, irregular heart beats and raised blood pressure. There is clear evidence that it improves endurance significantly, and its misuse goes along way back with athletes of old chewing raw coffee beans before events.*

## 2.5.1 ANAEROBIC GLYCOLYSIS

The initial stages in the breakdown of carbohydrates occur in the cytoplasm of all cells, including muscle fibres. Oxygen is not involved in these initial stages in the cytoplasm and they are therefore referred to as **anaerobic glycolysis**. Anaerobic glycolysis generates a net gain of two ATP molecules from each molecule of glucose, which only represents about 5% of the total ATP that can be generated from the complete breakdown of the glucose molecule. However, this process is important in supplying ATP in high intensity short term exercise, especially in fast twitch muscle fibres, as it supplies ATP more than twice as quickly as the aerobic system.

An essential part of these stages of anaerobic glycolysis, is the removal of hydrogen atoms from the substrate *(in a process known as oxidation)*. This hydrogen is accepted by the coenzyme NAD, which is thus reduced to NAD2H *(NAD with two hydrogen atoms)*. For anaerobic glycolysis to continue, unreduced NAD must be regenerated by the removal of the 2H *(oxidation)* from the NAD2H. This is achieved in different ways depending on the availability of oxygen.

## 2.5.2 LACTIC ACID METABOLISM

When the demand of the muscles for oxygen exceeds the supply, as for example at the beginning of most types of exercise, but particularly during intense efforts lasting up to about a minute, an intermediate product *(pyruvic acid)* of the breakdown of glucose, is used to oxidise the NAD2H. In doing this, the pyruvic acid itself becomes reduced to lactic acid. The complete breakdown of glucose is thus prevented, as an intermediate product of the breakdown of glucose is being used to form lactic acid, instead of going on to be oxidised to carbon dioxide and water with the release of much larger amounts of energy in the mitochondria.

A small amount of lactic acid is always formed in the muscles, even at rest or during the lightest exercise, but if the rate of clearance is more than or equal to the rate of production, lactate levels are held by homeostatic *(steady state)* systems at what can be called a '**lactate steady state**'. When the production does exceed clearance, lactate accumulates increasingly in the muscle and then consequently in the blood. High lactate concentrations *(and the resulting acidic effect)* sensitise nerve endings, which create an uncomfortable sensation either locally in the muscle or throughout the whole body, depending upon the mode of exercise and the type and number of motor units recruited. Ultimately, the effects of very high levels of lactate, and associated hydrogen ions ($H^+$), can act as a major factor in causing a person to stop exercising through an interference with the muscle contraction process *(see section 1.7)* and pain.

## Of Lactic & Lactate

An acid is a compound which dissociates in solution to release hydrogen ions $(H^+)$. The release of $H^+$ ions leaves the rest of the molecule with a net negative charge. This negatively charged part is known as the **lactate** ion.

Lactic acid $\rightarrow$ Lactate ion $+ H^+$

$$C_3H_6O_3 \rightarrow C_3H_5O_3^- + H^+$$

One of the major components of 'endurance' is the ability to perform high intensity workloads without lactate accumulating. A 'point' has been established beyond which further increases in work result in an accumulation of lactate. Many terms have been given to this 'point' and will be keenly debated for years to come, but is commonly accepted as the lactate threshold (LT) and less commonly now the anaerobic threshold. The point at which LT occurs is a key physiological determinant for performance in endurance events as it represents a parameter of sub-maximal exercise performance. Below the LT, oxygen uptake matches the requirement of aerobic respiration (i.e. steady state) and is not maximal. The upper limit of exercise intensity that can be sustained in a steady state manner, is known as the maximal lactate steady state. At this level blood lactate production transiently exceeds the rate of clearance, but although blood lactate concentration is elevated it remains stable.

Although, some consider the term LT misleading since anaerobic 'lactic' respiration occurs during all types of exercise *(even low intensity)* to a greater or lesser extent and a 'threshold' is not always clearly established, nevertheless, the term LT is becoming increasingly accepted by scientists, coaches and athletes. However, a more recent style of terminology may be more accurate, namely the **lactate 'breakpoint'**.

The most common method used to determine LT or the lactate 'breakpoint' by applied exercise physiologists is the subjective examination of blood lactate/work profiles. This method is open to individual interpretation, yet it allows a consideration of individual differences, which with experience can become consistently accurate.

There are normally between 0.5-2.0 mMs of lactate per dm$^3$ *(litre)* of blood at rest and during steady state exercise, and most often the 'breakpoint' will occur between 2-4 mM. However, the concentration at which the 'breakpoint' exists will vary greatly e.g. between 1-6 mM have been observed. Therefore the use of a fixed blood lactate concentration to represent the LT can be inaccurate. However, increasingly the term onset of blood lactate accumulation *(OBLA)* is becoming synonymous with the fixed lactate concentration of 4mM, since at those levels lactate is often 'accumulating', but this does not always apply and individual responses should be considered.

Regardless of the complications in analysis and terminology, the purpose of monitoring blood lactate response to exercise is to;

❖ establish the exercise intensity which LT occurs

❖ prescribe training intensies

❖ assess the effectiveness of a training programme when the objective is to improve the LT.

### ❖ Training and Performance Applications

The lactate threshold (LT) may be expressed as a measure of exercise intensity (i.e. watts or kilometres per hour) which provides an index of functional performance, and is often used to predict athletic performance, e.g. 10km run time. Alternatively, the point can be expressed as a percentage of the person's aerobic capacity ($\dot{V}O_2$ max) (see section 4.9), indicating the relative level that can be sustained in relation to a person's aerobic capacity. This concept is now referred to as **fractional utilisation** as it reflects the ability to 'utilise' (or maintain the use of) a 'fraction' (or percentage) of the aerobic capacity.

In untrained subjects the lactate threshold (LT) may occur between 50-60% of $\dot{V}O_2$ max, whereas in elite endurance athletes the LT may be at 90% of their $\dot{V}O_2$ max. These highly trained athletes possess a very high fractional utilisation, which will allow a high exercise intensity to be maintained without the lactate accumulating and causing fatigue.

Blood lactate levels in relation to intensity of effort in untrained and trained individuals.

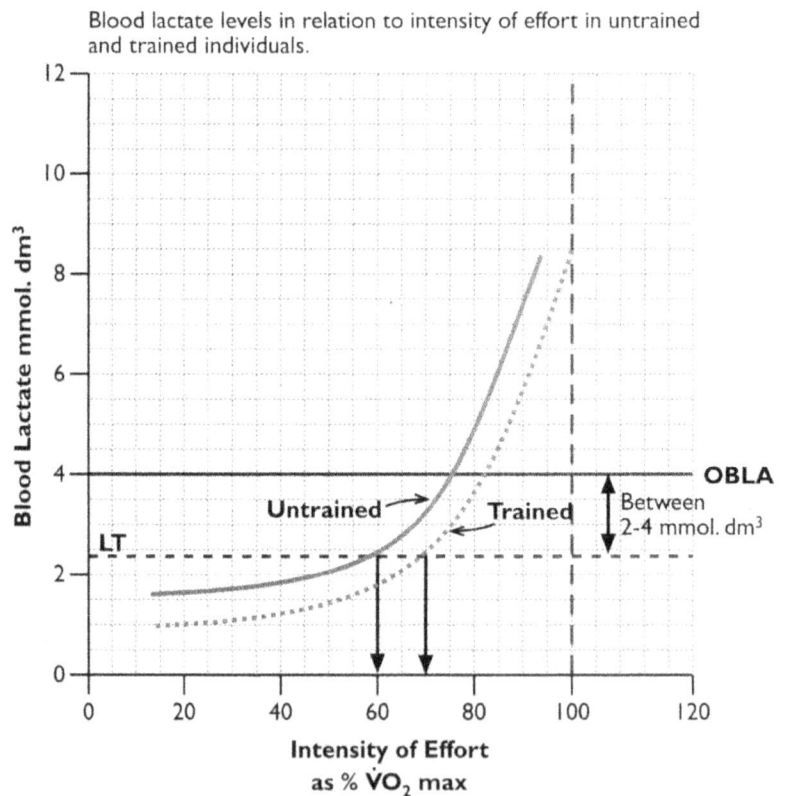

There are several ways by which lactic acid can be removed from the muscles. Some of them are of relatively minor importance, namely its loss in sweat and urine, and its reconversion to glycogen in the liver and in the muscles. This reconversion to glycogen in fact requires more energy than was yielded by the original anaerobic glycolysis, so that there is no overall gain of energy if this occurs.

The major pathway for the removal of lactic acid is its aerobic oxidation to carbon dioxide and water, yielding ATP. This can occur in the muscle fibres themselves once sufficient oxygen is available, especially in the slow twitch fibres of muscles not taking part in the acute exercise. Or the lactic acid can be transported in the blood to organs which are not operating under anaerobic conditions, where it can be oxidised, for example the heart, brain, liver, and kidneys. As it can be used in this way it is not therefore a waste product as such.

These processes are operating during exercise to a greater or lesser extent, and continue after exercise until the blood lactate concentration returns to resting levels. The removal of lactate is such that its concentration is halved about every 15 minutes.

### 2.5.3 AEROBIC PATHWAYS

If sufficient oxygen is delivered to the working muscles it can be used to re-oxidise the NAD2H produced in the oxidation of glucose to pyruvic acid. As described before, this results in the production of water and the regeneration of the NAD. Therefore, instead of being converted to lactic acid, the pyruvic acid is free to continue down the metabolic pathway that leads to. its complete oxidation into carbon dioxide and water, and the production of many more molecules of ATP.

These aerobic stages take place in the cell structures or organelles known as mitochondria. Slow twitch muscle fibres have large numbers of initochondria containing the enzymes, coenzymes, and cytochroine pigments necessary for the aerobic stages. All of these are increased as a result of aerobic training.

The complete oxidation of pyruvic acid occurs via a cyclic process known as the Krebs Cycle which produces more NAD2H. This NAD2H is oxidised back to NAD by the removal of hydrogen atoms, which are subsequently oxidised by oxygen, producing water, and releasing energy which is trapped in ATP.

This complete oxidation of a glucose molecule results in the production of 31 molecules of ATP.

**Figure 2.2:** *Simplified diagram of metabolic pathways involved in energy release from glycogen and glucose during and immediately after exercise*

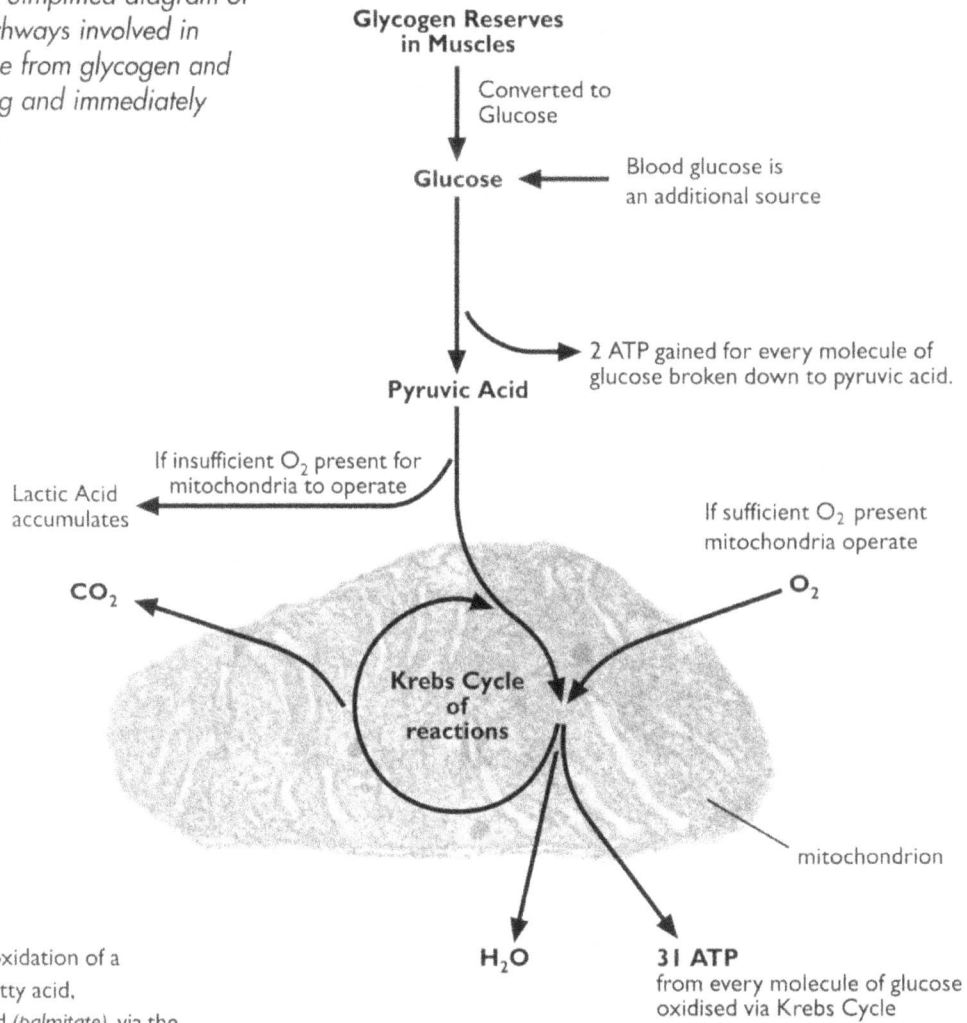

**Glycogen Reserves in Muscles**

Converted to Glucose

**Glucose** ← Blood glucose is an additional source

2 ATP gained for every molecule of glucose broken down to pyruvic acid.

**Pyruvic Acid**

If insufficient $O_2$ present for mitochondria to operate

Lactic Acid accumulates

If sufficient $O_2$ present mitochondria operate

$CO_2$

$O_2$

**Krebs Cycle of reactions**

mitochondrion

$H_2O$

**31 ATP**
from every molecule of glucose oxidised via Krebs Cycle

The complete oxidation of a molecule of a fatty acid, e.g. Palmitic Acid *(palmitate)*, via the Krebs Cycle yields 102 ATP.

**Summary Equation**

$$C_6H_{12}O_6 + 6O_2 \rightarrow 6CO_2 + 6H_2O + 31 \text{ ATP}$$

**Figure 2.3:** *Percentage contribution from each of the energy systems to 10, 30, 90 seconds of maximal exercise*

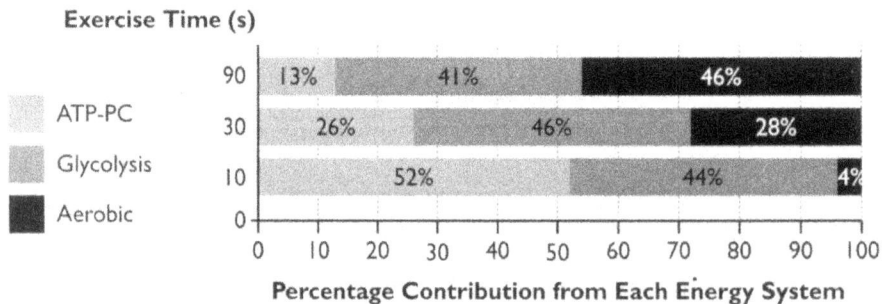

**Exercise Time (s)**

| | ATP-PC | Glycolysis | Aerobic |
|---|---|---|---|
| 90 | 13% | 41% | 46% |
| 30 | 26% | 46% | 28% |
| 10 | 52% | 44% | 4% |

**Percentage Contribution from Each Energy System**

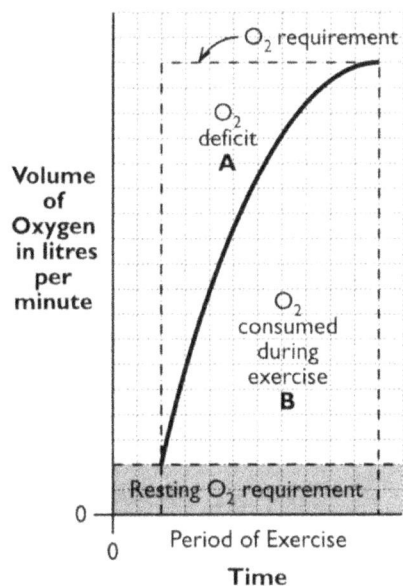

A + B = Total oxygen demand during period of exercise in excess of that required at rest.

## 2.6 OXYGEN REQUIREMENT FOR RECOVERY AFTER EXERCISE

### 2.6.1 OXYGEN DEFICIT

The oxygen deficit is defined as: the difference between, the calculated amount of oxygen necessary to supply the energy *(from aerobic respiration)* for a period of exercise *(therefore removing the need for any anaerobic respiration)*, and the actual amount of oxygen absorbed during the exercise period in excess of that used in an equivalent period of rest. In other words, the amount of oxygen that the subject was short of during the exercise.

### 2.6.2 OXYGEN DEBT

During recovery from exercise, the oxygen consumption remains high, above the rate of consumption for an equivalent period of rest. Originally this amount of extra oxygen required during recovery was assumed to be that which was needed for the removal of the accumulated lactic acid, and was therefore referred to as the 'oxygen debt'. However, there is only a slight relationship between the oxygen deficit and the so-called 'oxygen debt'; with the 'oxygen debt' being up to twice the oxygen deficit. In other words the 'oxygen debt' is not simply involved with the repayment of a deficit incurred during exercise. Therefore a better term is the excess post exercise oxygen consumption *(EPOC)*.

### 2.6.3 EXCESS POST EXERCISE OXYGEN CONSUMPTION (EPOC)

There are two components to the EPOC, namely the fast component and the slow component. In the fast component, the oxygen consumed is used to regenerate the phosphagen system of ATP and PC, and to resaturate the myoglobin and tissue fluids with oxygen. If the exercise was mainly aerobic this would be replaced within several minutes.

In recovery from more strenuous exercise, which has resulted in the accumulation of lactate, and raised the body temperature, there is an additional slow component, and it may be as long as 24 hours after stopping exercise before the oxygen consumption returns to pre-exercise levels.

Much oxygen is also used to supply many other processes which are not related to the deficit incurred during exercise. Oxygen is needed to supply the muscles of the thorax (chest) responsible for the increased breathing rate, and the more rapid beating of the heart, and also to maintain the higher metabolic rate due to the raised body temperature and the residual effects of the hormones adrenaline and thyroxine. The redistribution of ions (such as calcium, potassium and sodium) in the fluids inside and outside of cells after exercise, also requires energy and therefore oxygen.

**Figure 2.4:** *To show the relationship between oxygen deficit and EPOC*

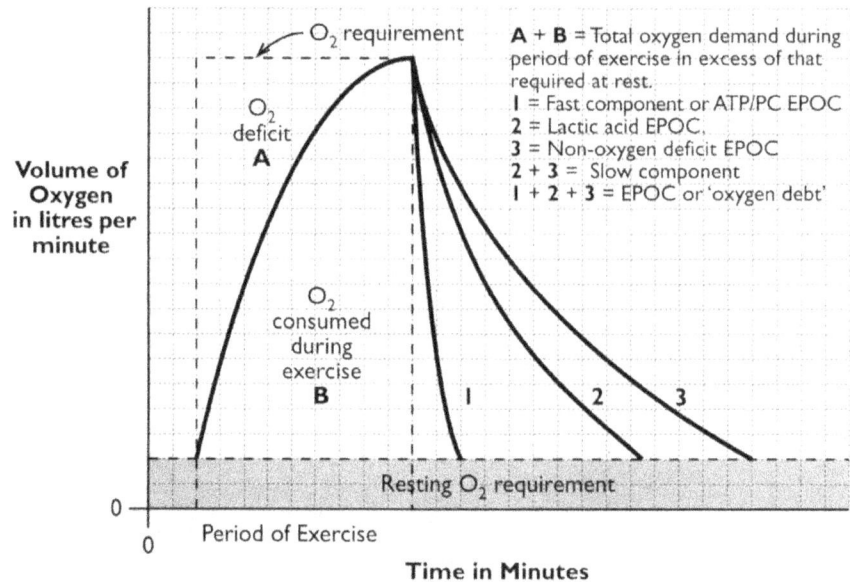

**2.7 ENERGY EFFICIENCY**

The energy released by the oxidation of energy-rich substrates in respiration can be estimated indirectly from the **net oxygen cost** of a particular steady state activity. The net oxygen cost being the amount of oxygen used in the activity, above that which would have been consumed in an equivalent period of rest *(B in fig. 2.4)*.

For example the oxidation of 180 g glucose requires 134.4 dm³ *(litres)* of oxygen, and releases 2867 kJ of energy.

Therefore, for a given volume of oxygen uptake the energy released can be calculated. However, the energy released for a given volume of oxygen varies, depending upon whether carbohydrates, fats, or proteins are oxidised.

The particular substrate being used can be estimated from the **respiratory exchange ratio** *(RER)* or **respiratory quotient** *(RQ)*, where

$$RER = \frac{\text{Volume of } CO_2 \text{ output per minute}}{\text{Volume of } O_2 \text{ uptake per minute}} = \frac{\dot{V}_{CO_2}}{\dot{V}_{O_2}}$$

RER *(fat and carbohydrate)* in relation to %$\dot{V}_{O_2}$ max *(after Astrand)*

As the intensity of effort increases RER increases indicating the decrease in the aerobic oxidation of fats, and the increasing aerobic and anaerobic oxidation of carbohydrates.

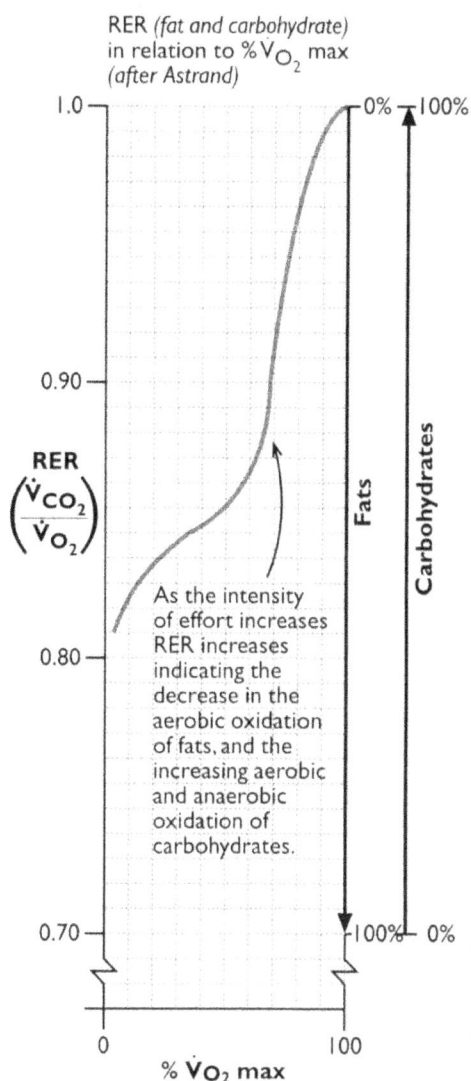

The particular substrate being used can be estimated from the **respiratory exchange ratio** *(RER)* or **respiratory quotient** *(RQ)*, where

$$RER = \frac{\text{Volume of } CO_2 \text{ output per minute}}{\text{Volume of } O_2 \text{ uptake per minute}} = \frac{\dot{V}_{CO_2}}{\dot{V}_{O_2}}$$

The values for carbohydrates, fats, and proteins are 1.0, 0.7, and 0.8 respectively. Intermediate figures result from the oxidation of mixtures of substrates, and any anaerobic respiration. Any anaerobic respiration increases the RER as $CO_2$ is produced with no $O_2$ uptake.

Having calculated the energy released from the net oxygen cost of a period of exercise, an attempt can be made to calculate the efficiency of the system, where :

$$\% \text{ Efficiency} = \frac{\text{Useful Work Output}}{\text{Energy Released}} \times 100$$

If the activity used in the measurement of useful work output is weight bearing, as on a treadmill, then the results should be calculated per unit body weight

There are inefficiencies in the trapping of energy in ATP, and in converting the chemical energy of ATP into useful mechanical work. So it is estimated that overall human movement is at best about 25% efficient with regard to energy use.

### ❖ *Training and Performance Applications*

*Maximum efficiency occurs at a specific optimum work rate, above which there can be a dramatic fall off in efficiency. So that in endurance events, operating at the optimum rate can lead to large energy savings, and therefore to greater improvements in performance than could be achieved by greater effort. This is particularly well illustrated in cycling, where the correct riding position and use of gears is of the utmost importance.*

### 2.7.1 GENERATION OF HEAT

The energy not trapped in ATP, and ultimately even the energy that is released from the ATP in doing work, is all lost as heat.

Therefore, the amount of energy released from the oxidation of respiratory substrates, is equal to the amount of heat given off by the body. As this occurs mainly over the skin, the results should be calculated per unit surface area of the body.

The energy released as heat is not wasted, it enables the body to maintain a core body temperature of 37°C, which provides optimum conditions for the enzyme controlled activities of metabolism.

## 2.8 HEAT AND FLUID BALANCE DURING EXERCISE

If the body temperature rises above about 37°C, then the blood vessels in the dermis of the skin shunt more blood towards the surface, increasing the amount of heat lost by radiation, convection, and conduction to the surroundings, if the external temperature is lower than body temperature.

If the core body temperature rises to 0.5°C above normal, sweating begins. Energy is required to change the state of water from liquid to vapour, and this is known as the latent heat of evaporation. When sweat evaporates from the skin it absorbs this heat energy from the body, lowering the body temperature. One dm³ (litre) of sweat removes about 2424.4 kJ (579 Calories) of energy from the body. Under conditions of high relative humidity sweat may not evaporate, and if it accumulates on the skin it has no cooling effect.

### ❖ Training and Performance Applications

*During exercise more heat is generated than at rest, and this speeds up the enzyme controlled processes involved in muscle contraction, which in turn aids exercise. This is one reason why a 'warm-up' period improves performance. However, if the body temperature rises above a certain critical limit, temperature control mechanisms come into operation. If these cannot cope, the body temperature may rise causing a drop in performance; if the temperature continues to rise above a certain critical level 'heat stroke' can occur which may be fatal.*

### 2.8.1 WATER LOSS

During exercise there can be large water losses via sweating. The volume of water lost depends on the external temperature and humidity, the amount of exercise, and the volume of water drunk before and during exercise. Losses can vary from 1 - 3 dm³ (litres) per hour. This can lead to decreased plasma volume and consequent circulatory problems, such as decreased stroke volume and increased heart rate, there may also be reduced blood pressure and reduced tissue fluid formation. Also, even minor dehydration impairs muscle function. With more severe dehydration the increased concentration of salts in the plasma can interfere with normal heart function, leading to possible heart failure. In addition several hours of sweating can lead to sweat gland fatigue.

At high exercise intensities there can be competition for blood between the active muscles and the skin, and the flow to the skin can be reduced in favour of maintaining the blood flow to the active muscles. This will reduce the heat loss from the skin.

### ❖ *Training and Performance Applications*

*Training increases the functional capacity and sensitivity of the sweating response, with sweating starting at a lower body temperature, and larger volumes of more dilute sweat being secreted, than in untrained subjects. If weight loss due to dehydration is greater than 2% of body weight after a particularly hard session, then further training should not be undertaken until the losses are made good. At a critical point of water loss during exercise sweating stops, to maintain the blood volume, and the core body temperature rises dramatically with heat stroke occurring at around 40.5°C. Fluid and energy losses are best replenished by drinking a cold (8-13°C) 2.5% solution of complex carbohydrates, such as maltodextrins from partially broken down starch. Such drinks empty from the stomach as fast as pure water, and go some way to replacing energy losses. Recommended amounts are, 500 cm³ 30 minutes before exercise commences, and 150 cm³ at 15 minute intervals during exercise, especially in events lasting longer than 60 minutes. The subjective sense of 'thirst' is not an accurate indicator of the amount of water needed, so drinks should be taken as a matter of course, rather than only when feeling thirsty.*

Salts are also lost in the sweat, the most important being sodium chloride. Generally there is a lower sodium concentration in the sweat than in the plasma, therefore sweating does not decrease the plasma sodium concentration. However, if the fluid losses are replaced by drinking pure water, the plasma sodium level does decrease, and this is implicated in muscle cramps. Also as a result of this decrease in concentration of the fluids which surround the cells, water is drawn by osmosis into the cells, causing swelling of the tissues.

## SUMMARY

❖ Glycerol, fatty acids, glucose, and amino acids can all be used as respiratory substrates, and have metabolic pathways in common.

❖ Fats are the main respiratory substrate in aerobic respiration. They cannot be broken down anaerobically. For a given volume of oxygen fats yield less ATP than carbohydrates.

❖ Carbohydrates become more important as an energy source as the intensity of exercise increases, because they are the only substrates that can be broken down anaerobically as well as aerobically. However, their supplies are limited.

❖ Proteins yield amino acids. Excess amino acids cannot be stored in the body, they are deaminated in the liver to produce urea, and the remainder of the molecule enters the respiratory pathway to be oxidised with the release of energy.

❖ The hormones adrenaline, noradrenaline and glucagon increase the availability of energy for exercise. Insulin decreases the availability of energy for exercise.

❖ There are three components of the energy supply system, ATP/PC, anaerobic, glycolytic, and aerobic. All are involved in all exercise to varying degrees.

❖ The phosphagen system of ATP and PC represents the most immediately available source of energy for muscles, especially in powerful contractions. ATP can be resynthesised using energy from PC, anaerobic respiration and aerobic respiration. PC can only be resynthesised using energy from ATP.

❖ Anaerobic respiration generates little ATP and results in the production of lactic acid, which contributes to fatigue and breathlessness.

❖ The point at which the intensity of exercise results in the accumulation of lactate in the blood is known as the 'lactate threshold' *(LT)*.

❖ Aerobic respiration generates much more ATP, and results in the complete oxidation of respiratory substrates, for example glucose, to carbon dioxide and water.

❖ The amount of oxygen one is short from the onset of exercise *(thus requiring anaerobic respiration to occur)* is known as the oxygen deficit.

❖ The concept of oxygen 'debt' is not simply involved with the 'repayment' of an oxygen deficit incurred during exercise. Therefore a better term is the 'excess post exercise oxygen consumption' *(EPOC)*.

❖ The efficiency of energy usage can be calculated from the net oxygen cost of a particular activity and the useful work output. With maximum efficiency occurring at an optimum work rate.

❖ The substrate being used in respiration can be estimated from the respiratory exchange ratio *(RER)* or respiratory quotient *(RQ)*.

❖ All energy is ultimately lost as heat.

❖ Heat loss is by convection, conduction, radiation and sweating. At high exercise intensities blood is withdrawn from the skin as it is shunted to the working muscles, thus reducing heat loss.

❖ Sweating results in water and salt loss, which can have a critical effect on the efficiency of exercise.

# 3 CIRCULATION IN ACTION

## OBJECTIVES

To enable the reader to understand the following.

❖ The structure and function of the heart.

❖ The cardiac output as a product of the heart rate and stroke volume.

❖ The ways in which the heart adapts to exercise in both the short and long term.

❖ Blood pressure and its regulation with respect to exercise.

❖ The role of blood vessels in the circulation of the blood during exercise.

❖ The formation of tissue fluid and the role of the lymphatic system.

## 3.1 THE STRUCTURE OF THE HEART

The heart is a four chambered muscular pump which contracts rhythmically to force the blood around the body. One-way valves ensure that the blood flows through the heart in the correct direction

**Figure 3.1:** *The Human heart - internal structure. Note that deoxygenated blood from the body flows from the venae cavae into the right atrium, into the right ventricle, and to the lungs in the pulmonary arteries. Oxygenated blood returns from the lungs in the pulmonary veins into the left atrium, into the left ventricle, and to the body in the main aorta.*

**Anatomical Drawing of Heart**

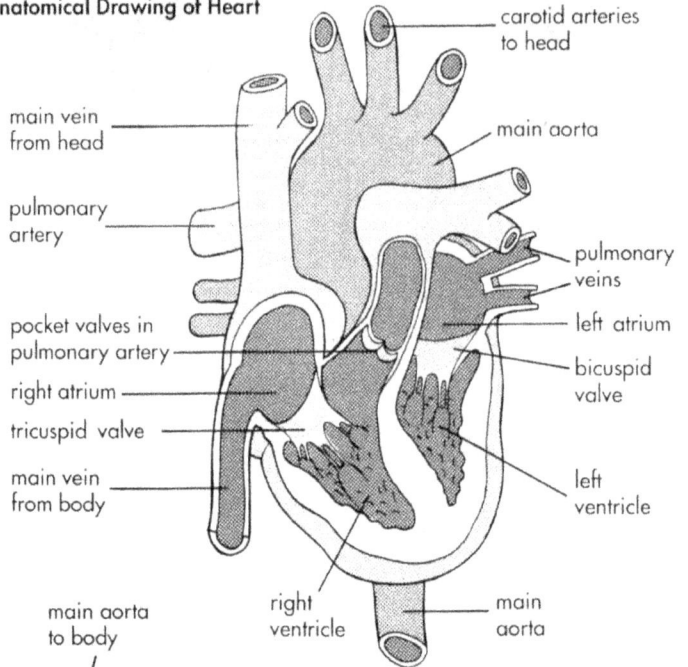

carotid arteries to head

main vein from head

main aorta

pulmonary artery

pulmonary veins

pocket valves in pulmonary artery

left atrium

bicuspid valve

right atrium

tricuspid valve

main vein from body

left ventricle

right ventricle

main aorta

**Simplified Heart Diagram**

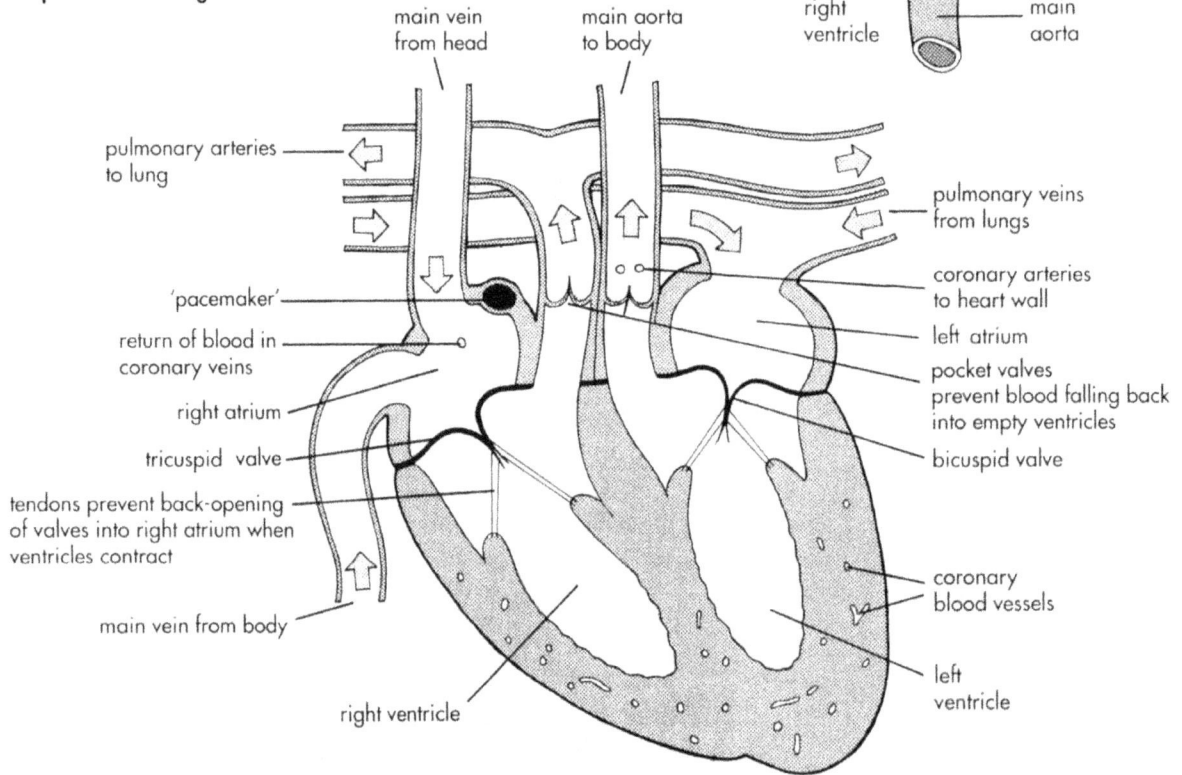

main vein from head

main aorta to body

pulmonary arteries to lung

pulmonary veins from lungs

'pacemaker'

coronary arteries to heart wall

return of blood in coronary veins

left atrium

right atrium

pocket valves prevent blood falling back into empty ventricles

tricuspid valve

bicuspid valve

tendons prevent back-opening of valves into right atrium when ventricles contract

coronary blood vessels

main vein from body

left ventricle

right ventricle

Double circulation, the blood passes through the heart twice in each complete circulation of the body.

**Basic Circulation**

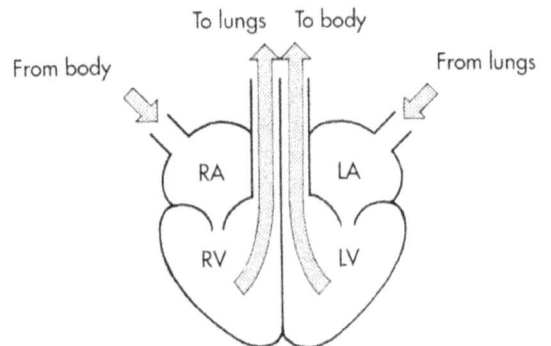

To lungs    To body

From body

From lungs

RA    LA

RV    LV

**Figure 3.2:** *Diagrammatic representation of the human circulation system.*

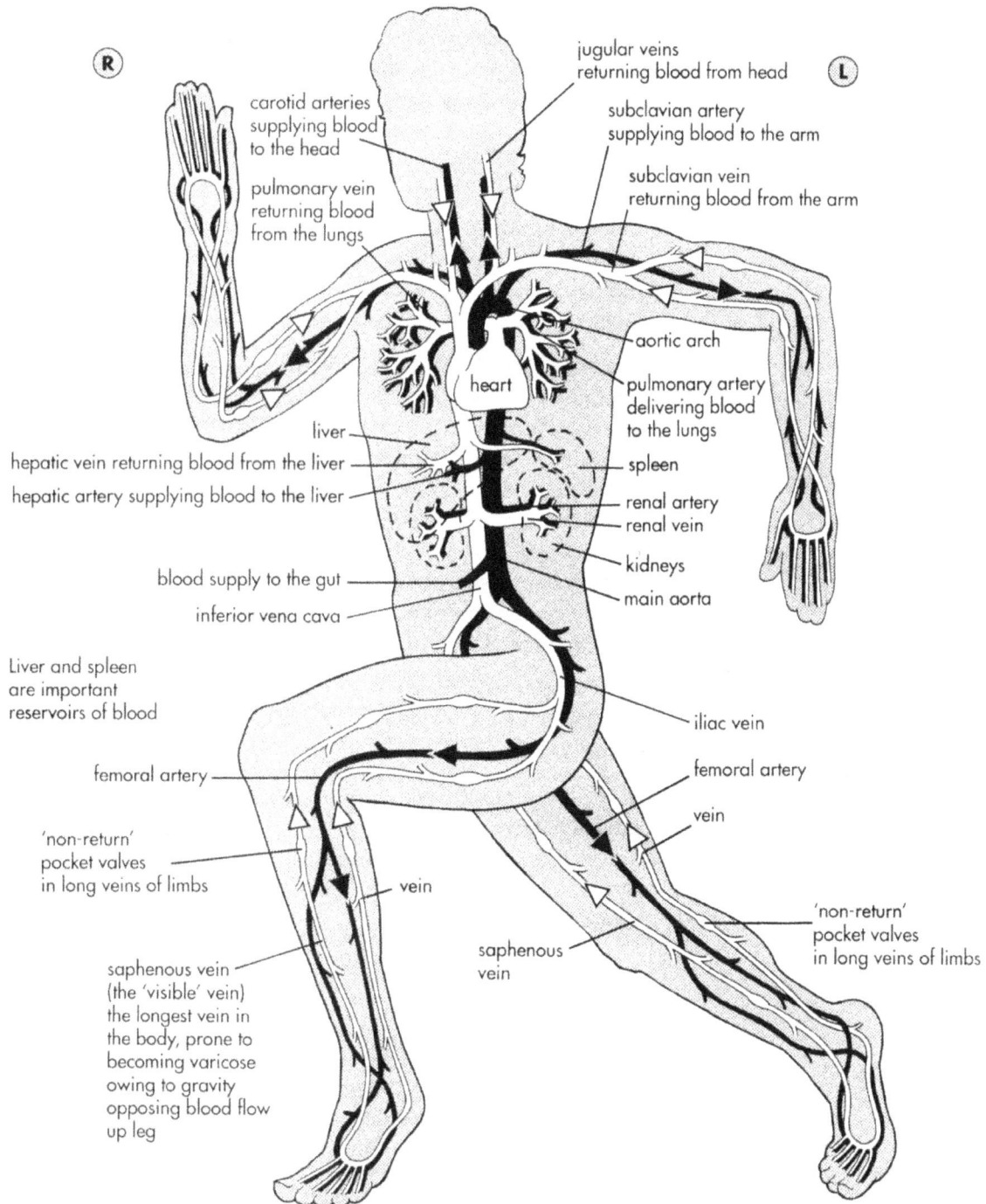

jugular veins
returning blood from head

subclavian artery
supplying blood to the arm

subclavian vein
returning blood from the arm

carotid arteries
supplying blood
to the head

pulmonary vein
returning blood
from the lungs

aortic arch

pulmonary artery
delivering blood
to the lungs

heart

liver

spleen

hepatic vein returning blood from the liver

hepatic artery supplying blood to the liver

renal artery
renal vein

kidneys

blood supply to the gut

main aorta

inferior vena cava

Liver and spleen
are important
reservoirs of blood

iliac vein

femoral artery

femoral artery

vein

'non-return'
pocket valves
in long veins of limbs

vein

'non-return'
pocket valves
in long veins of limbs

saphenous
vein

saphenous vein
(the 'visible' vein)
the longest vein in
the body, prone to
becoming varicose
owing to gravity
opposing blood flow
up leg

R    L

**Figure 3.3:** *The cardiac cycle.*

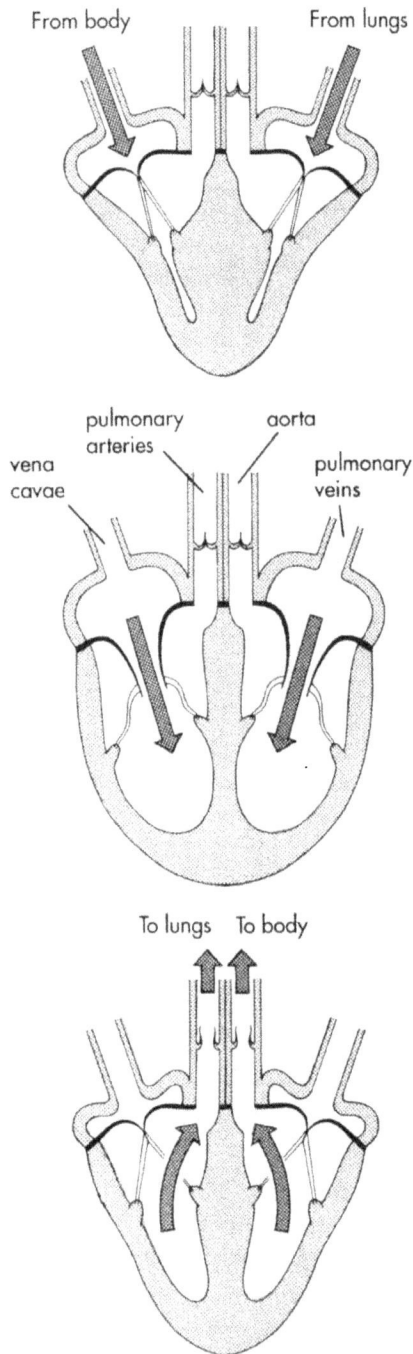

From body     From lungs

pulmonary     aorta
arteries
vena                  pulmonary
cavae                 veins

To lungs   To body

## 3.2 THE FUNCTION OF THE HEART

### 3.2.1 CARDIAC CYCLE

The sequence of events that occurs during the filling and emptying of the heart is known as the cardiac cycle.

At a rate of 70 beats per minute (bpm) the complete cycle takes about 0.86 seconds. It is a continuous sequence of events, but for the purposes of explanation it is convenient to consider it as occurring in a series of stages

There is a point in the cardiac cycle when all the heart valves are shut, which can be taken as the starting point.

Blood returning under low pressure in the main veins *(superior and inferior venae cavae)* from the upper and lower body enters the right atrium, and at the same time blood returning from the lungs in the pulmonary veins enters the left atrium.

The atria fill with blood, and the rising pressure of the blood in the atria pushes the tricuspid and bicuspid valves open, and both ventricles begin to fill. As the heart fills with blood both atria contract *(atrial systole)* forcing even more blood into the ventricles. These are now stretched, and they both contract *(ventricular systole)* forcing the blood upwards. This upsurge of blood forces the tricuspid and bicuspid valves shut. These valves are prevented from back-opening into the atria by tough cords, which are held taut by contraction of the papillary muscles to which they are attached. The closing of these valves makes the first heart sound.

The blood is thus forced along the pulmonary artery to the lungs, and along the main aorta to the rest of the body, forcing open both sets of pocket or semi-lunar valves on the way.

When the ventricles relax *(ventricular diastole)* the blood tends to fall back down the pulmonary artery and main aorta under gravity, thus filling and shutting the pocket valves. The closing of these valves makes the second heart sound. All valves are now shut and the cycle is repeated.

### 3.2.2 CARDIAC MUSCLE

The muscle tissue of the heart wall is a special type known as cardiac muscle. Cardiac muscle is composed of short striated fibres interconnected by side branches, which allow for the rapid spread of excitation from fibre to fibre throughout the heart wall.

Cardiac muscle has a long refractory or rest period *(during which it cannot be restimulated)*, so that no merging or summation can occur between successive contractions. This ensures that the heart as a whole beats in a rhythmic fashion, with a period of rest or diastole between contractions, during which it can refill. The rest period also protects the cardiac muscle from fatigue.

Cardiac muscle cannot respire anaerobically to any significant extent, and being aerobic it uses glucose, fatty acids and lactic acid *(formed in the exercising muscles)* as respiratory substrates or fuels.

Of all tissues, cardiac muscle has the highest concentration of mitochondria *(organelles within cells which are the centres of aerobic respiration)*, and each fibre is supplied with at least one capillary. Therefore cardiac muscle has a high capability for aerobic respiration *(oxidative potential)*, being able to extract up to 80% of the oxygen from the blood even at rest *(compared to as little as 25% extraction by other tissues)*. As a result of this high extraction of oxygen at rest, the increased demand for oxygen during exercise can only be supplied by an increase in coronary blood flow.

### 3.2.3 CORONARY BLOOD SUPPLY

The cardiac muscle of the heart wall is supplied with blood via the coronary arteries. These arise from the main aorta, just above the pocket valves. In this position their supply of blood is continuous, irrespective of the state of contraction of the heart. There is less flow in the coronary arteries during systole as they are compressed by the contraction of the cardiac muscle. As the heart rate increases during exercise, diastole shortens more than systole. This would be expected to reduce the diastolic coronary blood flow. However, there is still an increase in coronary flow from about 250 cm$^3$ per minute at rest to 1000 cm$^3$ per minute during exercise, as a result of increased dilation of the coronary blood vessels which is stimulated by hypoxia *(a drop in the oxygen level in the cardiac muscle)*, and by increased aortic pressure which forces more blood into the vessels.

---

❖ *Training and Performance Applications*

*Aerobic training increases the stroke volume, and therefore also lowers the resting heart rate, thus increasing the period of diastole in each cardiac cycle, and allowing greater coronary blood flow. A low heart rate and a high stroke volume is the most efficient way of achieving a given cardiac output, and as a result of this, the cardiac muscle of the trained heart needs less oxygen and therefore a proportionally lower coronary blood supply at rest than the untrained heart.*

*As a result of the special properties of cardiac muscle, the heart cannot develop an oxygen debt to any significant extent, nor can it develop 'cramp', as can occur in other muscle tissue.*

*It is generally agreed that a healthy heart cannot be damaged or 'strained' by exercise, no matter how strenuous. However if cardiac muscle is deprived of oxygen for a only a few minutes then it will die. This is what happens if the coronary blood vessels become blocked. Fatty deposits or plaques of cholesterol on the walls of the coronary arteries narrow the arteries and roughen their internal surface, these changes can trigger the formation of an internal blood clot or thrombus (coronary thrombosis) which may further block the artery. Regular aerobic training reduces the risk of such events, but unfortunately does not offer complete protection to those at risk.*

### 3.2.4 CARDIAC OUTPUT ($\dot{Q}$)

The cardiac output is the amount of blood pumped into the arteries by contraction of the ventricles in a given time. At rest it may be about 5000 cm$^3$ *(5 dm$^3$)* per minute from each ventricle in an untrained individual, rising to about 30 000 cm$^3$ per minute from each ventricle in a trained individual undergoing maximal exercise

The cardiac output is the product of two components, the **heart rate**, and the **stroke volume**. The stroke volume is the volume of blood pumped out by a single contraction of the ventricles, typically about 75 cm$^3$ by each ventricle when the body is at rest.

Cardiac Output *(dm$^3$ per minute)* =
Heart Rate *(beats per min)* × Stroke Volume *(dm$^3$)*

The **cardiac reserve** is the difference between the cardiac output at rest *($\dot{Q}$ rest)* and the maximum cardiac output *($\dot{Q}$ max)*. It gives a measure of the increase in blood supply available during exercise.

Cardiac Reserve dm$^3$.min$^{-1}$ = $\dot{Q}$max − $\dot{Q}$rest

### 3.2.5 HEART RATE

Cardiac muscle is **myogenic**, that is it can contracts without nervous stimulation. Co-ordinating waves of electrical activity originate in a mass of specialised cardiac muscle in the right atrium, the sinu-atrial node *(SA node or 'pacemaker')*. This provides the basic rhythm of the heart contractions.

The wave of excitation spreads rapidly through the interconnected cardiac muscle fibres of the atria. However a band of fibrous connective tissue between the atria and ventricles prevents the wave of excitation spreading down into the ventricles. Another mass of specialised tissue, the atrial-ventricular node at the bottom of the atria is stimulated, and impulses are transmitted to the base of the ventricles in a tract of conducting tissue, from where the impulses radiate upwards through the cardiac muscle of the ventricles.

This pattern of the spread of excitatory waves through the heart, ensures that the atria contract to force the blood down into the ventricles, and that the ventricles contract to force the blood up into the pulmonary arteries and main aorta. The spread of the wave of excitation throughout the heart can be detected by electrodes attached to the skin of the chest and displayed as an electrocardiograph *(ECG trace)*.

The onset of strenuous activity can result in ECG abnormalities due to insufficient blood flow to the heart muscle in the first few seconds. This can be prevented by a two minute warm up period before exercise.

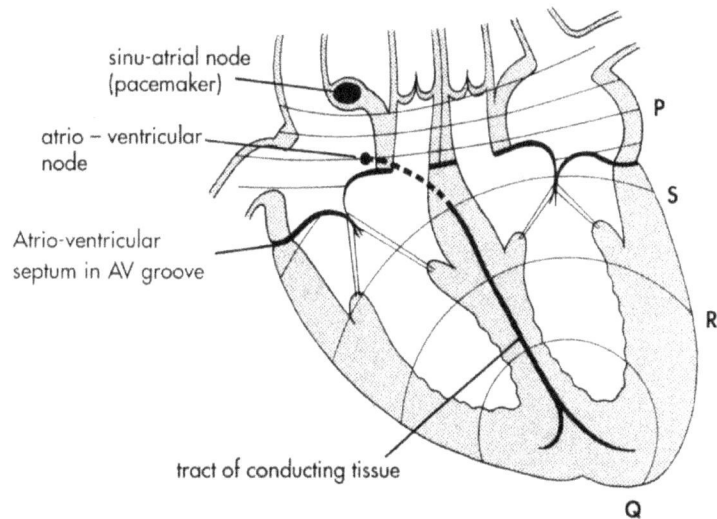

**Figure 3.4:** *Electrical co-ordination of the contraction of the heart and ECG trace.*

sinu-atrial node (pacemaker)

atrio – ventricular node

Atrio-ventricular septum in AV groove

tract of conducting tissue

P

S

R

Q

ECG trace of electrical activity of heart

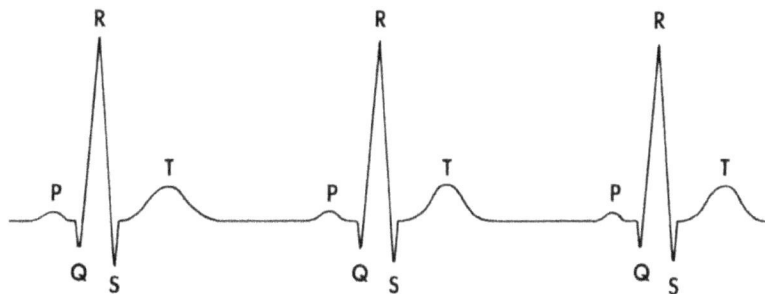

**P** - *corresponds with the spread of the excitation wave over the atria. This positive deflection precedes atrial systole.*

**QRS** - *corresponds to the conduction of the excitation wave through the ventricles.*

**T** - *corresponds to the positive deflection which accompanies the relaxation of the ventricles.*

The basic rate of contraction due to the spontaneous rhythm of the sinu-atrial node, can be altered by a variety of influences, over a range from about 30 bpm to as many as 220 bpm.

The heart is influenced by the activity of the **sympathetic** and **parasympathetic** nerves of the autonomic nervous system, which originate from the cardiovascular control centre in the brain.

The sympathetic nerves secrete **noradrenaline**, which stimulates an increase in the heart rate *(tachycardia)* directly. Also, in response to a general activation of the sympathetic nervous system, **adrenaline** is released from the adrenal glands, which also increases the heart rate.

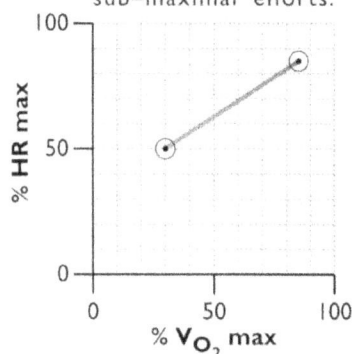

Linear relationship between heart rate and oxygen uptake at sub–maximal efforts.

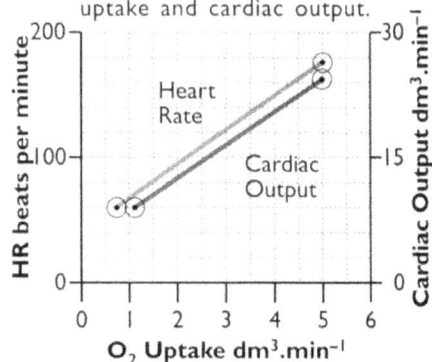

Linear relationship between $O_2$ uptake and heart rate, $O_2$ uptake and cardiac output.

### ❖ Training and Performance Applications

*Aerobic endurance training produces a lower resting heart rate. Endurance athletes typically (but not necessarily) have a resting heart rate of 50 bpm or lower. Although a reduced resting heart rate (bradycardia) is associated with good endurance capacity, it is not necessarily a good predictor of performance. The maximum heart rate cannot be increased, and in fact it may be reduced as a result of training due to the increased stroke volume. It is very difficult for some endurance trained athletes to raise their heart rate to the 180 bpm (See p 113) required in theory by some traditional interval training sessions.*

The parasympathetic nerves, release **acetylcholine** which causes a decrease in the heart rate *(bradycardia)*.

During exercise, the progressive increase in the heart rate is due at first to a decrease in parasympathetic activity, and then to the increasing activity of the sympathetic nervous system. These alterations in the nervous stimulation of the heart are themselves the result of changes associated with exercise acting either directly or indirectly on the cardiovascular control centre in the brain. These changes include the following.

An accumulation of carbon dioxide and lactic acid. A decrease in the amount of oxygen in the blood. An increase in temperature, which speeds up metabolism in general *(and which also reduces the viscosity of the blood making it easier to pump)*, and nervous reflexes from sensory receptors in the working muscles and joints feeding back information about the increased mechanical activity associated with exercise. In addition an increase in the filling of the heart *(as a result of a greater venous return, which in turn occurs as a result of the increased activity of the muscle and respiratory pumps)*, causes stretching of the SA node, which increases its rhythm of nervous discharge

The heart rate increases very rapidly at the onset of exercise, and in a trained athlete it can treble within a minute of starting exercise. This increase is too rapid to be the result of the activities of the metabolic mechanisms concerned with cardiac acceleration described above, and therefore the exact cause is not clear. However, connections exist between the higher centres of the cerebral cortex, and the cardiovascular centre of the brain *(from which the sympathetic and parasympathetic impulses to the heart originate)* which can result in anticipatory increases in heart rate prior to exercise.

The increased heart rate helps in the uptake and delivery of more oxygen to the tissues, to meet their metabolic demands.

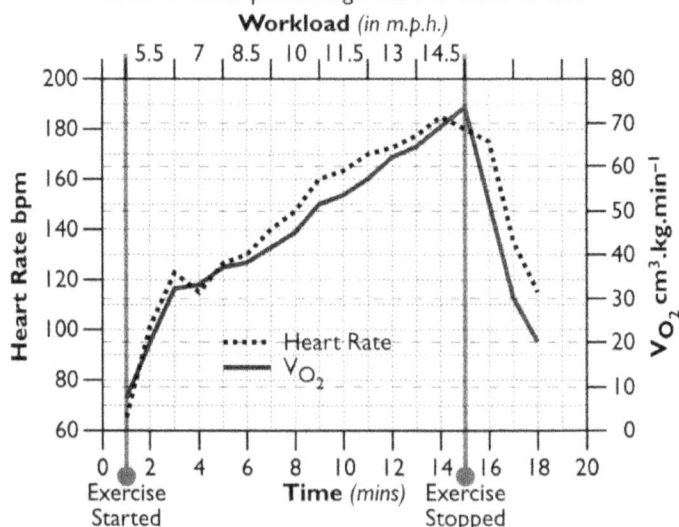

Graph to show heart rate and oxygen uptake of a middle distance runner performing a maximal treadmill test.

Changes in heart rate, stroke volume, and cardiac output with increasing workloads.

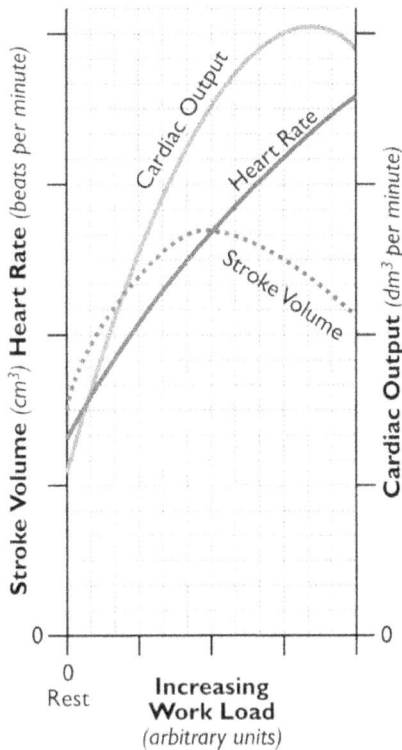

### 3.2.6. STROKE VOLUME

When the body is not exercising, the ventricles do not empty completely when they contract, they only expel about 40-70% of the blood that they contain, this is known as the ejection fraction.

With the onset of exercise; adrenaline and noradrenaline increase the contractility of the cardiac muscle. This in turn increases the force of contraction and therefore the ejection fraction, which is the main factor in an increased stroke volume.

Starling's law states that the force of contraction of the ventricles is proportional to the degree of stretch of the muscle fibres during filling. The reduced heart rate in the trained individual allows a greater filling during the longer diastole, so the degree of stretch of the cardiac muscle fibres will be greater. This in turn increases the ejection fraction, and therefore the stroke volume.

Heart rate has been shown to increase proportionally with increasing workload, whereas stroke volume does not. It has been shown that stroke volume increases to a maximum *(plateau)* between 40-60% of $\dot{V}O_2$max. However, in conflict, it is suggested that in highly trained athletes stroke volume continues to increase beyond this level to maximum effort, perhaps because of adaptations to training.

❖ *Training and Performance **Applications***

*Over a period of aerobic training the heart enlarges, as a result of both an increase in muscle mass and size of the chambers, thus also increasing the stroke volume. For a given cardiac output, as the stroke volume increases the heart rate decreases. For example at rest, the demands on the circulation of the trained and the untrained individual are the same, so their cardiac outputs are also the same, but the trained subject achieves that cardiac output with a slower heart rate as a result of their increased stroke volume. A decreased heart rate for a given cardiac output means that the heart is having to work less hard, and requires less oxygen. Endurance trained athletes initially increase their cardiac output in response to the demands of exercise mainly by an increase in their ejection fraction, whereas untrained individuals respond mainly with an increase in heart rate. Thus in trained endurance athletes the heart rate increases more slowly during exercise. For periods of more than about 30 minutes of sub-maximal work, the cardiac output is maintained, but due to loss of fluid as sweat, there is a reduced plasma volume and therefore reduced venous return to the heart, therefore the stroke volume gradually decreases and the heart rate gradually increases, in a process known as 'cardiovascular drift'. The heart rate also increases in proportion with the increase in motor unit recruitment as exercise progresses.*

It is important to remember that the heart responds to the **demands** of the body, especially the skeletal muscles, whose demand for oxygen can increase up to twenty times during exercise. The heart is therefore responding to feedback information from the tissues. In this way the cardiac output is adjusted to maintain the necessary blood supply under the correct pressure to the tissues.

A greater cardiac output requires an equivalent increase in the return of blood to the heart. This venous return of blood to the heart is affected by the position of the body. For example during upright exercise such as running there are difficulties in increasing venous return against the force of gravity, so the central blood volume is decreased, reducing the venous return and the stroke volume, and resulting in an increase of the heart rate for a given cardiac output. In the horizontal position venous return is not opposed by gravity, and therefore during exercise, such as swimming, there are less problems in ensuring adequate venous return, the central blood volume and venous return are not decreased, therefore compared to the situation in running the stroke volume is greater and the heart rate slower for a given cardiac output.

## 3.3 ATHLETE'S HEART

The adaptations that the heart shows to aerobic exercise over a period of time, lead to the condition known as 'athlete's heart'.

There is enlargement *(hypertrophy)* of the heart giving an increased stroke volume and a reduced resting heart rate, which is often irregular and associated with abnormal electrical activity *(ECG)*. This abnormal activity can result in misdiagnoses if the examining physician is unaware of the training background.

There is increased capillarisation of the cardiac muscle, but it is not clear whether this is due to the development of new capillaries, or the more complete use of existing ones.

The individual myofibrils in the cardiac muscle fibres thicken and increase in number, and there is an increase in the size and number of the mitochondria.

The healthy untrained heart does not experience insufficient oxygen supply *(hypoxia)* even during maximum exercise, but the adaptations seen in 'athletes heart' are an advantage, as they enable the heart to function at a lower percentage of its total oxidative capacity during exercise.

❖ *Training and Performance Applications*

*Training induced changes of the heart may be considered as either* **functional**, *or* **structural**. *Functional changes, such as an increase in contractility of the cardiac muscle, are rapid and occur within weeks; but structural changes, such as the enlargement of the ventricles takes months or even years. These changes are reversed over a period of time once the training load is decreased, which is a contributory cause of the general reversibility of the training effect, as a result of which the level of performance returns to the pre-training level.*

## 3.4 BLOOD PRESSURE

The heart pumps blood around a complete circuit of vessels, which offer resistance to the flow of blood, therefore pressures are generated within the system. In a healthy adult, the pressure in the arteries close to the heart is highest *(120 millimetres (mm) of mercury (Hg))* when the ventricles are contracting, which is known as the **systolic blood pressure**, and lowest *(80 mm Hg)* when the ventricles are relaxing, which is known as the **diastolic blood pressure**.

The average of the systolic and diastolic pressures during a complete cardiac cycle, known as the **mean blood pressure**, determines the overall rate of blood flow through the circulation.

The blood pressure drops as the blood gets further away from the left ventricle and circulates around the body. Some typical figures for the mean blood pressure *(in mm Hg)* whilst at rest are: arteries 100, arterioles 60, capillaries 18, veins 7, and at the entrance to the right atrium 3.

Blood pressure is determined by the **cardiac output** and **the resistance to flow of the blood.** The resistance is often referred to as the peripheral resistance, as the major part of it occurs in the smaller vessels *(mainly the arterioles)* at a distance from the heart. The resistance to flow is due to the friction between the blood and the walls of the blood vessels. The friction is determined by the length, diameter, and the smoothness of the lining of the vessel; and the viscosity of the blood.

Arteries and arterioles *(and to a lesser extent veins and venules)* have involuntary muscle fibres in their walls, which when contracted constrict the vessel reducing its diameter. Also the diameter of the vessel is effectively reduced by any loss of elasticity of the vessel walls, such as occurs with age and/or the formation of fatty deposits *(plaques)* of cholesterol. The plaques also roughen the lining of the vessels which further increases the friction between the blood and the walls.

The viscosity of the blood is increased if the plasma volume is decreased as a result of dehydration, for example through sweating during prolonged exercise. Replacement of fluid by drinking at regular intervals during prolonged exercise is essential to maintain water and salt balance, and temperature regulation *(Chapter 2 Section 2.8, p 54)*.

The viscosity of the blood also increases if the number of red blood cells *(RBC)* in a given volume of blood is increased. This can occur as a result of general endurance training, but particularly at altitude *(Chapter 4 Section 4.5, p 93)*, as a response to the shortage of oxygen.

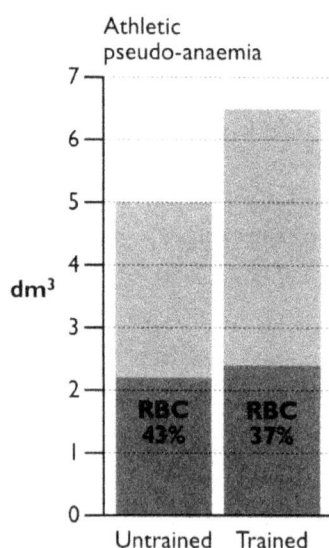

Athletic pseudo-anaemia — bar chart showing dm³ for Untrained (RBC 43%) and Trained (RBC 37%).

### ❖ *Training and performance applications*

*The viscosity of the blood is reduced after taking aspirin, thus aiding circulation and lowering blood pressure, it is also an analgesic, dulling pain. Whether these effects have any significant influence on sport performance is not clear, and the practice of taking aspirin as an ergogenic aid (one that improves the ability to do work) is not banned. Indeed a very well known distance runner takes an aspirin a day as a matter of routine. Endurance training can increase the overall blood volume by up to 10%, mainly as a result of an increase in the amount of plasma, and this would also tend to reduce the viscosity of the blood. The increase in the amount of plasma in endurance trained athletes effectively reduces the concentration of haemoglobin in the blood, a condition known as 'athletic pseudo-anaemia', even though the number of red blood cells is increased.*

In the horizontal position the pressure along an artery is highest closest to the heart; but when vertical, gravity causes pooling of blood in the legs, so that arterial and venous pressures are raised in the lower parts of the legs. This can lead to swelling of the feet and ankles over long periods of time, so that allowance must be made in the size of footwear in events involving long periods on the feet. The pooling of blood in the lower legs also interferes with the return of blood in the veins to the heart. During rhythmic exercise, the cardiac output rises, and there is a rise in the systolic blood pressure, perhaps by as much as 30%. This may gradually decrease as the peripheral resistance is reduced by an increased opening up of blood vessels in working muscles as exercise continues. The diastolic pressure, however, remains relatively constant throughout such exercise. At the same relative work load, both the systolic and diastolic blood pressures increase more when the work is carried out with the arms, rather than the legs.

When the peripheral resistance is high, as for example when the arterioles lose their elasticity as a result of age, the pressure generated by the cardiac output at systole stays high throughout diastole, even though at that point there is no cardiac output. Regular exercise over a period of time can result in a lowering of the resting systolic and diastolic blood pressures in those with such raised blood pressure, possibly as a result of a decrease in the peripheral resistance as the circulation becomes more efficient.

The pulmonary circulation to the lungs operates under a lower pressure than the main systemic circulation of the body, for example the pulmonary arterial systolic blood pressure is between 19 and 26 mm Hg. The resistance is lowered by the thin elastic walls of the pulmonary arteries, and the relative thinness of the muscle walls of the arterioles. This lower pressure is necessary to prevent the exudation of tissue fluid into the alveoli, which would flood the lungs and interfere with the uptake of oxygen.

---

### ❖ *Training and Performance Applications*

*The systolic and diastolic blood pressures are raised considerably during static (isometric), straining type exercises, as in weight training. This is caused by compression of the peripheral blood vessels, causing an increase in the resistance to blood flow. The increase in blood pressure is even greater if the effort is made whilst holding the breath. So weightlifters can experience large rises in blood pressure when lifting, especially when only using the arms. This increased blood pressure, however, in turn helps to overcome the compression of the peripheral blood vessels, and maintain the circulation to the contracted muscles.*

---

### 3.4.1 CONTROL OF BLOOD PRESSURE

Despite all the factors acting to alter the blood pressure, various mechanisms operate to maintain the blood pressure within optimum limits for as long as possible.

Pressure receptors in the aorta, carotid arteries, and the heart, monitor the blood pressure at all times, and feed sensory information to the vasomotor and cardiac centres in the brain. Reflex actions occur to oppose any move away from the optimum.

If the blood pressure begins to drop, there is an increased activity of the sympathetic nervous system. This increases the heart rate and stroke volume, and therefore the cardiac output. It also causes constriction of the arterioles and the veins, which increases the peripheral resistance. These changes result in a raising of the blood pressure back to normal. At the same time there is an an increase in the release of the hormones adrenaline and noradrenaline.

Adrenaline causes constriction of the arterioles *(vasoconstriction)* in the skin and the gut, but dilation of the vessels in the skeletal muscles, the liver, and the coronary circulation of the heart. It also increases the heart rate and cardiac output. Thus also contributing to the rise in blood pressure.

Noradrenaline causes a net vasoconstriction overall, even in the skeletal muscles. As a result noradrenaline has a stronger effect in raising blood pressure than adrenaline.

In addition, the adrenal glands secrete the hormone aldosterone, which causes the kidneys to retain salts and water. This increases the blood volume and therefore the blood pressure.

If the blood pressure begins to rise, there is a reflex increase in the parasympathetic nerve impulses, which causes the heart rate to slow, decreasing the cardiac output. The blood vessels are also dilated which lowers the peripheral resistance, thus further reducing the blood pressure.

During exercise, although initially both the cardiac output and blood pressure increase, these various homeostatic mechanisms act to restrict the rise in blood pressure, and to eventually bring it down.

## 3.5 ROLE OF VESSELS IN THE CIRCULATION OF THE BLOOD

The blood circulates around a continuous system of blood vessels, the arteries, arterioles, capillaries, venules and veins.

### 3.5.1 ARTERIES

The arteries close to the heart have a large cross sectional area, and thick elastic walls. The large elastic arteries are stretched by the cardiac output of ventricular systole, and they recoil when the arterial blood pressure drops during ventricular diastole. This recoil assists the circulation of the blood around the body, helps smooth the flow between contractions of the heart, and helps to maintain the blood pressure. This stretching and recoiling of the arteries is felt as **the pulse**.

### 3.5.2 ARTERIOLES

Further from the heart, the arteries branch into the smaller arterioles. The arterioles penetrate all tissues of the body and connect to beds of finely branching capillaries. There are so many capillaries in the body, that they cannot all be supplied with blood at the same time. Therefore there is competition for blood between different regions of the body, especially during exercise, when blood must be **shunted** to the working muscles, and consequently withdrawn from other regions. For example, when the body is at rest, about 45% of the cardiac output passes through the capillaries of the gut wall and associated glands, and the kidneys. During maximum exercise this can be reduced to about 3% of the cardiac output, as blood is redirected to the working muscles. This does not necessarily lead to as large a reduction in oxygen supply to the gut and kidneys as it would seem, as these regions only use about 10 - 25% of the oxygen available in their normal blood supply. The reduced flow rate is therefore compensated for by a greater extraction of oxygen from the blood, by these tissues. However, prolonged reduction in the blood flow to the gut and its associated glands will inevitably interfere with digestion. In short intense periods of exercise the blood flow to the skin is also reduced, even in hot conditions, as it is shunted to the muscles. However, it is increased in longer duration moderate exercise in order to promote heat loss as part of temperature control.

The shunting of blood between competing tissues is achieved by constriction and dilation of the arterioles, and of the arteriovenous vessels, which are direct connections between the arterioles and venules, from which the capillary beds arise. The entrances to the capillary beds are controlled by circular precapillary sphincter muscles, which when contracted shut off the capillaries, and when relaxed open them.

**Table 3.1:** *Estimated blood flow in cm³ per minute, to different organs/systems in a trained male at rest and during maximum effort.*

Estimated blood flow in cm³ per minute

| Organ system | At rest | % | Max. effort* | % |
|---|---|---|---|---|
| Skeletal muscle | 1000 | 20 | 26 000 | 88.00 |
| Coronary vessels | 250 | 5 | 1200 | 4.00 |
| Skin | 500 | 10 | 750 | 2.50 |
| Kidneys | 1000 | 20 | 300 | 1.00 |
| Liver & gut | 1250 | 25 | 375 | 1.25 |
| Brain | 750 | 15 | 750 | 2.50 |
| Whole body | 5000 | 100 | 30 000 | 100.00 |

*Cycle ergometer

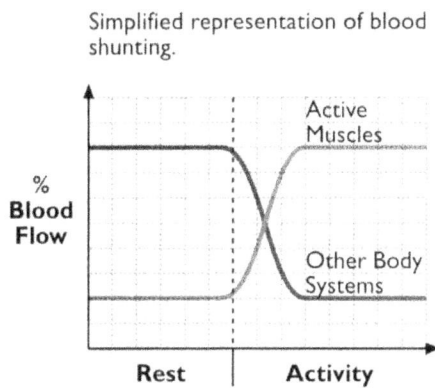

Simplified representation of blood shunting.

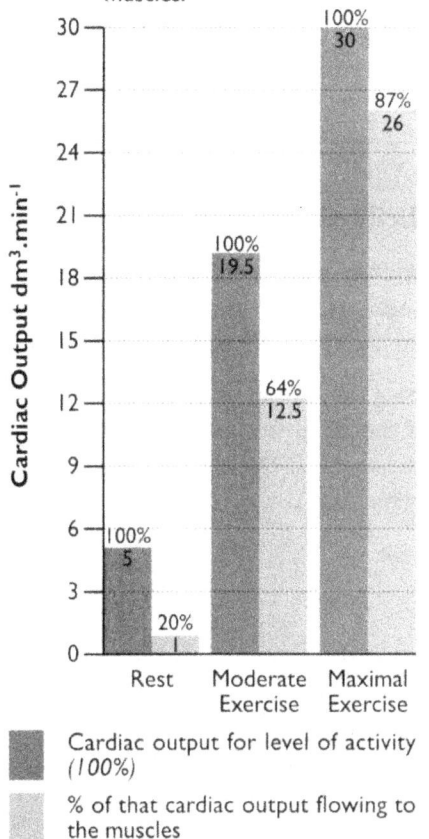

Effect of exercise on cardiac output and redistribution of blood to muscles.

Cardiac output for level of activity (100%)

% of that cardiac output flowing to the muscles

**Figure 3.6:** *Relationship of capillary beds and through vessels*

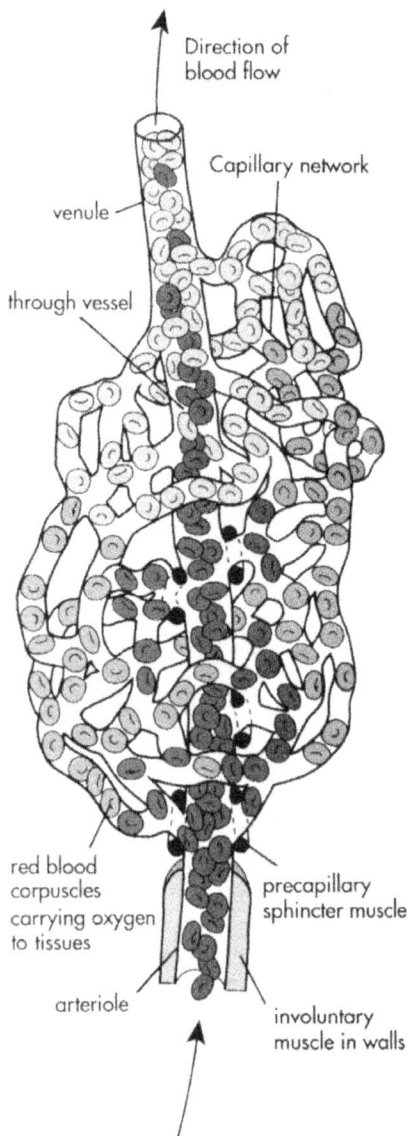

❖ *Training and Performance Application*

*As the withdrawal of blood from the gut to supply active muscles inevitably interferes with digestion, vigorous exercise should not be undertaken until at least three hours after eating a meal. Very long distance endurance performances require some form of energy intake during the event, as is seen in cycle racing, long distance swimming and ultra distance running, but measures are taken to keep the 'digestion load' to a minimum by taking in readily digestible and absorbable food. The reduction of blood flow to the liver and kidneys during exercise may be a contributory factor to fatigue, as these organs are involved in the removal of waste products produced during exercise. The removal of blood from the liver can cause the liver to shrink, stimulating pain receptors in the covering membranes, which is one explanation of the 'stitch'. Another explanation of the 'stitch' is that it is caused as a result of cramp or spasm of various muscles in the chest and/or abdomen, such as the diaphragm, between the ribs, or in the stomach wall. Alternatively it is suggested that stretching of the membranes holding the abdominal organs in place, especially when running where there is a 'bouncing action' could be the cause. This last explanation is supported by the observation that the 'stitch' is relatively rare amongst cyclists and swimmers with their smoother non-jarring action, and that 70% of all stitches occur on the right hand side where the liver, the heaviest organ in the abdominal cavity is suspended beneath the diaphragm. It is suggested that by breathing out when the left foot hits the ground, and breathing in when the right foot hits the ground, the diaphragm will be moving down with the liver at the point of right foot impact, thus lessening the risk and/or effects of a stitch.*

### 3.5.3 CONTROL OF BLOOD SHUNTING

All the arterioles, and the precapillary sphincters which control the entrance of blood into the capillary networks, are supplied with vasoconstrictor nerves from the sympathetic nervous system.

Stimulation by impulses from these nerves causes vasoconstriction *(a narrowing of the diameter)* of the arterioles, and contraction of the precapillary sphincters which closes off the capillary beds, reducing the blood flow to the regions that they supply.

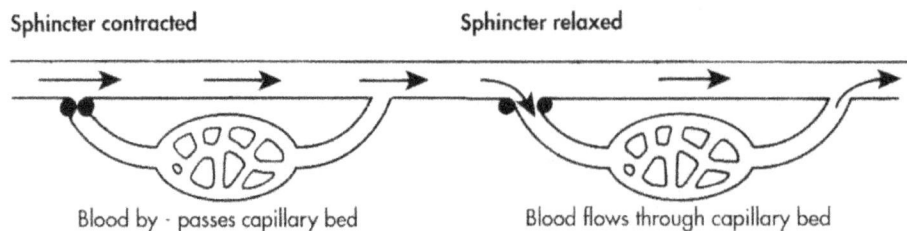

In addition, the adrenal glands secrete adrenaline and noradrenaline which cause the effects described earlier *(Section 3.4.1, p 69)*. However, during exercise this hormonal influence is less important than the faster sympathetic control.

The blood vessels supplying the cardiac muscle of the heart, and the skeletal muscles, are also supplied with sympathetic vasodilator nerves, which cause the muscle in the walls of the vessels to relax and the vessels to dilate, increasing the blood supply to the regions they supply. However, the major influences on vasodilation in the muscles are the local factors generated in the exercising muscles themselves. These local factors within the working muscle include a drop in the oxygen, and a rise in the carbon dioxide levels; a decrease in pH *(increase in acidity)*; an increase in temperature; an increase in the amount of ADP present in the muscle fibres; and an increase in potassium and magnesium ions in the muscles. These factors act directly on the involuntary muscle of the small arterioles, arterio-venous vessels, and precapillary sphincters, causing vasodilation. This effect is virtually instantaneous, and being controlled by products of muscle activity, is finely attuned to supplying the exact requirements of the working muscles. These local factors also stimulate highly sensitive sensory nerve endings which trigger the sympathetic vasodilation reflexes.

### 3.5.4 CAPILLARIES

The capillaries form a fine network which penetrate all tissues. They have the smallest diameter of all blood vessels, about 7 micrometres *(0.007 mm)*, which is the same as that of the red blood corpuscles. Therefore the corpuscles are forced to move slowly through the capillaries in single file, allowing the corpuscles to come close to the surrounding tissues, ensuring a short diffusion distance between them. Capillary walls are just one cell thick, which further aids exchanges by diffusion between the blood and the tissues. Also the huge number of capillaries provide a very large surface area across which these exchanges occur. As more capillaries open up in the working muscles, the surface area between the blood and tissues is increased, and the diffusion distance between the two is decreased, resulting in more rapid and efficient exchanges between them. For example total oxygen uptake by the muscles can increase up to twenty times during exercise.

### 3.5.5 TISSUE FLUID

As a result of the positive pressure of the blood, plasma is forced out through the walls of the capillaries to form tissue fluid. This tissue fluid carries oxygen, nutrients, hormones and antibodies, from the capillaries out into direct contact with the cells of the tissues. During heavy exercise as much as 10-20% of the plasma volume can be exuded as tissue fluid in the working muscles.

Most of the water in the tissue fluid moves back into the capillaries by osmosis, this occurs closer to the venules, where the blood pressure is lower. The remainder of the tissue fluid is drained away from the tissues in the blind-ending lymphatic capillaries of the lymphatic system, in which the fluid is known as lymph. As the fluid is drained from the tissues, it carries away waste products, as well as unused useful materials.

---

### ❖ Training and Performance Applications

*One cause of muscle pain and 'stiffness' after unaccustomed exercise is the 'waterlogging' of the muscles by excess tissue fluid which was not drained away efficiently during exercise by the lymphatic system. This 'waterlogging' causes the muscles to swell, which stimulates pain receptors. Training increases the efficiency of the lymphatic system in draining the tissue fluid away, so that the problem of muscle pain of this origin is reduced.*

---

### 3.5.6 VEINS

The venules from the capillary beds join together to form veins, which return the blood to the heart. The individual veins have a larger cross section area than the comparable arteries, and as there are more veins than arteries, the veins also have a greater total cross sectional area. The walls of the veins are thinner and less elastic than those of the arteries, and in the veins of the legs and arms there are semi-lunar or pocket valves at intervals along their length. These valves oppose any back flow of blood under gravity, and ensure one-way flow of blood to the heart. The thin walled veins with their large cross section present little resistance to the flow of the blood, and although after the blood has passed through the capillaries, the blood pressure in the veins is low, it still helps move the blood in the veins back to the heart. Because the blood in the veins is under low pressure, and their walls are thin, the veins are easily compressed by the smallest movements of the surrounding structures, particularly the skeletal muscles. As the muscles contract and relax during exercise they exert a massaging effect on the deep veins, especially for example in the legs, when running, swimming, or cycling. This massage effect of the 'muscle pump' is random and in no particular direction, but the pocket valves in these veins ensure that the flow is one-way, back to the heart.

In addition, breathing movements, known as the 'respiratory pump', which typically become more exaggerated during exercise, assist the return of blood in the veins to the heart. On breathing in, the diaphragm contracts and moves downwards, this decreases the pressure within the thorax, and increases the pressure within the abdomen. These changes in pressure compress the large veins in the abdomen, and assist the flow of blood into the veins of the thorax which return blood to the heart.

At rest the veins act as a large reservoir of blood for use when circulatory demands increase. Veins have a thin layer of involuntary muscle in their walls, contraction of which is controlled by sympathetic nerves, and the hormones adrenaline and noradrenaline. During exercise when the cardiac output increases, the muscle fibres in the walls of the veins are stimulated to contract (*venous tone*), the veins constrict, and a larger proportion of the venous blood is shunted back to the heart, especially from the veins in the legs and the abdomen. This is especially important in preventing blood from 'pooling' in the legs under the force of gravity during upright exercise, and immediately afterwards.

In these ways the venous return of blood balances the rising cardiac output. This is essential as the heart would be unable to maintain a high cardiac output unless it was receiving an equal volume of blood back from the veins.

### ❖ *Training and Performance Applications*

*After exercise, a 'warm-down' period of active movement is recommended, so that the 'muscle-pump' and to a lesser extent the 'respiratory pump', continue to massage the blood in the veins back to the heart. These, along with the venous tone, will prevent accumulation of blood in the legs, and therefore inadequate venous return, and consequent reduced cardiac output. Reduced cardiac output can lead to a decrease in the flow of blood to the brain and the possibility of fainting. If fainting occurs for this reason, it is self correcting, as the effect of gravity is removed in the horizontal position, and blood flow to the brain is restored. A 'warm-down' period also maintains the blood flow through the muscle capillaries, so that oxygen supplies are kept up, and carbon dioxide and lactic acid are removed; in addition the draining of tissue fluid away from the muscles in the lymphatic system is maintained.*

## 3.6 BLOOD VELOCITY

The velocity of the blood flow is related to the force with which it is pumped, and to the cross-sectional area of the vessels it is passing through. That is, the smaller the cross-sectional area, the faster the flow, and vice versa. Although the capillaries are the smallest blood vessels in the body, their total cross-sectional area is about 800 times greater than the aorta leaving the heart (*which has a diameter of about 25 mm).* Therefore the blood slows considerably in the capillaries, which allows for efficient exchanges with the tissues.

During exercise, the flow rate through the capillaries is increased as the cardiac output increases more rapidly than the number of extra capillary beds brought into action.

The blood speeds up on leaving the capillaries as a result of the total cross-sectional area of the veins being less than that of the capillaries. Within the venous system the blood speeds up as a result of the action of the 'muscle pump' and the 'respiratory pump', and as a result of the total cross sectional area of the venous system decreasing as venules join to form veins, and veins run into the main venae cavae which return blood to the heart.

Close to the heart the velocity of the blood in the veins is similar to that of the blood in the main arteries leaving the heart. This is necessary to ensure adequate venous return to balance the cardiac output. The flow in the pulmonary circulation around the lungs, is faster than that in the general *(systemic)* circulation, because the total cross-sectional area of the vessels in the pulmonary circulation is less than that in the systemic circulation. Note that velocity and pressure should not be confused. It is possible to have a low pressure with a high velocity, as for example in the veins.

**Figure 3.6:** *Relationship between blood pressure, velocity and total cross sectional area of blood vessels*

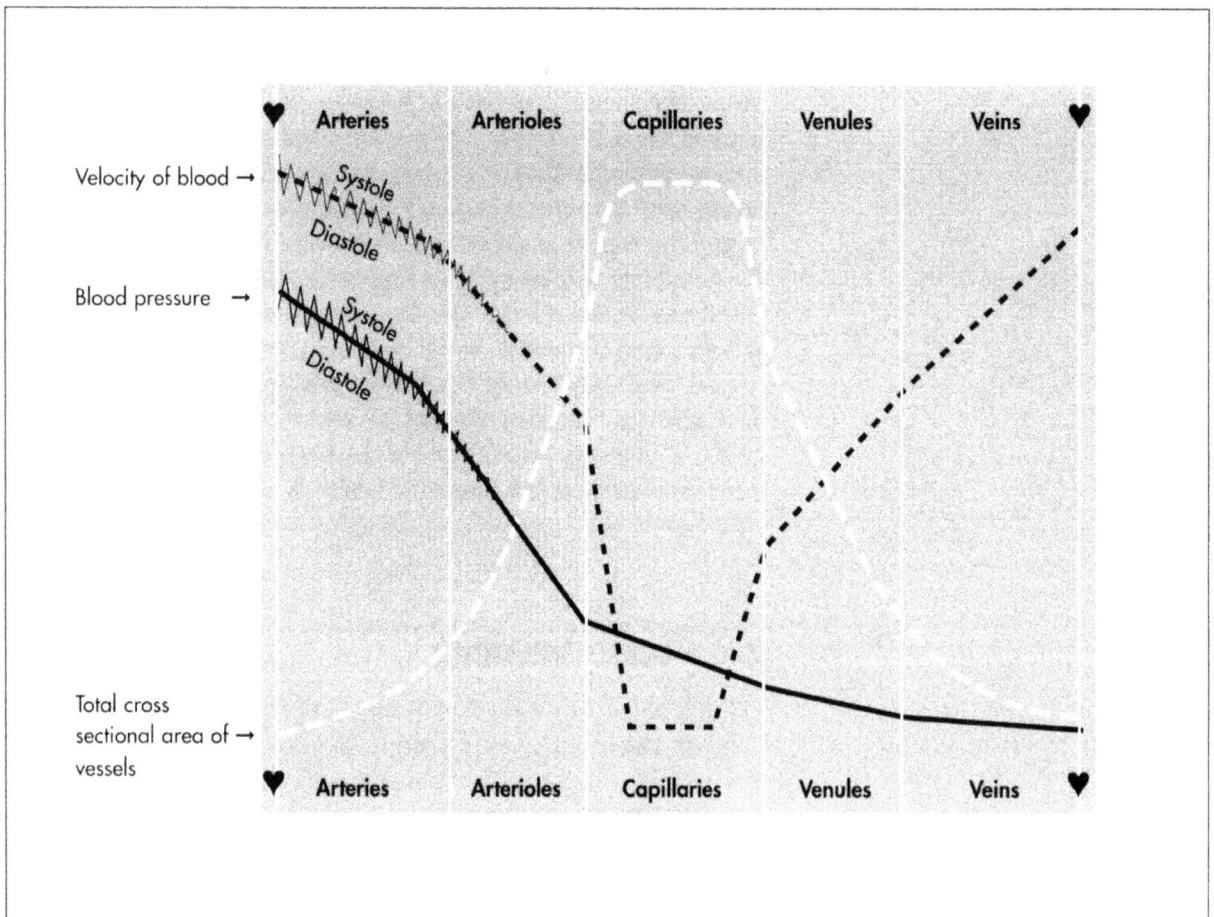

In conclusion it should be remembered that, taking all the complex interacting systems involved in the body's adaptation to the demands of exercise into consideration, it is the ability of the cardiovascular system to supply oxygen to the working muscles that largely determines aerobic endurance capacity.

## 3.7 THE LYMPHATIC SYSTEM

The lymphatic system is a system of vessels that drain excess fluid away from the tissues, back into the main veins near the heart.

Blind ending lymphatic capillaries are found draining all the tissues, and these join to form larger lymphatic vessels. Like the veins, these are thin walled, with numerous one-way pocket valves. The 'respiratory pump' and the 'muscle pump' move the lymph *(as the fluid is now known)* along the vessels to the re-entry point with the main veins.

At intervals in the larger vessels, there are swellings known as **lymph nodes**. These contain a sponge-like tissue with a large internal surface area, through which the lymph must pass. These lymph nodes contain special cells and antibodies which prevent any infective agents, which may have entered the tissues at vulnerable points *(such as the lining of the gut, the skin, the upper respiratory tract, or the lungs)* from entering the circulation.

Lymph is also important in returning plasma proteins to the blood. Some plasma proteins pass into the tissue fluid, and these proteins cannot pass back into the blood capillaries, but they can enter, and be removed by, the lymphatic capillaries. If these proteins accumulated in the tissues, they would cause retention of water, which can lead to painful swelling.

During exercise, more tissue fluid is exuded, but the increased breathing rate, and actively contracting muscles, cause the lymph to be drained away more quickly. Therefore the tissue fluid moves more rapidly across the tissues, increasing the amount of substances being transported to and from the muscle fibres during exercise.

## SUMMARY

❖ The right ventricle of the heart pumps blood to the lungs, and the left ventricle pumps blood around the body.

❖ The cardiac muscle of the heart wall has its own blood supply via the coronary blood vessels. Blood flows more easily through these vessels when the heart is relaxed, therefore the slower the heart rate, the greater the coronary blood supply.

❖ Cardiac muscle has a high rate of oxygen extraction from the blood, even when the body is at rest, therefore its increased demand for oxygen during exercise, can only be met by an increase in the heart's blood supply in the coronary arteries.

❖ Cardiac Output = Heart Rate per minute × Stroke Volume

❖ Cardiac muscle is myogenic, initiating its own contractions, but the heart rate is controlled by the pacemaker in the right atrium, which is influenced by a number of factors.

❖ The stroke volume increases during exercise. Endurance training increases the size of the heart, increasing the stroke volume even at rest. Therefore at rest the heart rate decreases *(bradycardia)*. Resistance training increases the thickness of the ventricle walls, but not the internal volume.

❖ To maintain a certain cardiac output, the same volume of blood must return to the heart.

❖ Blood Pressure = Cardiac Output × Peripheral Resistance.

❖ There are two measures of blood pressure, the higher taken when the heart is contracting *(systolic blood pressure)*, and the lower when the heart is relaxing *(diastolic blood pressure)*.

❖ Blood vessels play a major role in assisting the circulation of the blood. Arteries have muscular and elastic walls which expand and contract in *(the pulse)* response to the heart contractions helping to push the blood around the circulation. Veins have thin walls and are easily massaged by movement of surrounding muscles, one-way pocket valves ensure unidirectional flow back to the heart.

❖ Blood is shunted to different parts of the body according to demand. This is achieved by opening and closing capillary beds, especially in the skeletal muscles, the gut, and the skin.

❖ A 'warm down' period after exercise prevents accumulation of blood in the legs, ensures adequate venous return, and therefore enables the cardiac output to be maintained.

❖ The velocity of blood flow is not directly related to blood pressure. In the arteries both the pressure and velocity are high, however in the veins the pressure is low but the velocity is high.

❖ The lymphatic system returns excess tissue fluid to the blood circulation.

# 4 BREATHING GAS EXCHANGE AND TRANSPORT SYSTEMS IN ACTION

## OBJECTIVES

To enable the reader to understand the following.

❖ The structure of the thorax and lungs.

❖ Breathing movements, and their control.

❖ The use of ventilation as a marker of anaerobic threshold.

❖ Lung volumes and capacities.

❖ Gas exchange at the lung surface.

❖ The ways in which oxygen is transported from the lungs to the tissues.

❖ Gas exchange at the tissues.

❖ The ways in which carbon dioxide is transported from the tissues to the lungs.

❖ The concept of maximal oxygen consumption $(\dot{V}_{O_2}max)$.

## 4.1 STRUCTURE OF THE THORAX AND LUNGS

The thorax or chest cavity is an air tight cavity bounded by the rib cage, and the muscular sheet known as the diaphragm. It contains the heart and the lungs; and major blood vessels and the oesophagus *(food tube leading to the stomach)* pass through it. The air tubes *(trachea, bronchi, and bronchioles)* lead into the lungs, which are composed of millions of microscopic air sacs called alveoli. The lungs are held by 'suction' of the pleural fluid *(in the pleural cavity between the pleural membranes)* to the inside of the wall of the thorax, and being highly elastic in nature, expand and contract in response to the breathing movements of the thorax. The thorax is divided into two separate halves by a membrane, so that if one side is punctured only one lung will collapse.

**Figure 4.1:** *The Human thorax and lungs.*

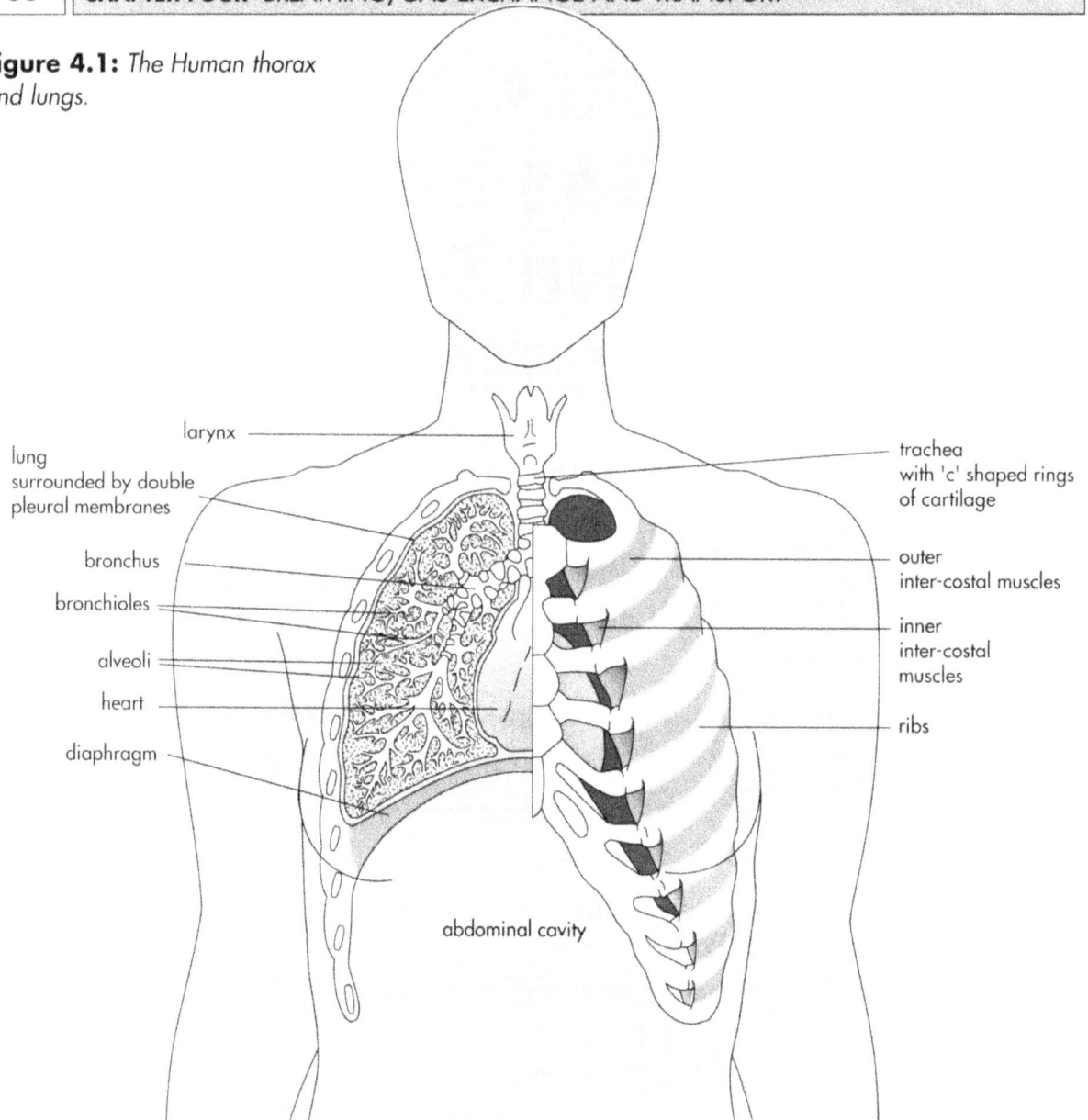

larynx

lung surrounded by double pleural membranes

bronchus

bronchioles

alveoli

heart

diaphragm

trachea with 'c' shaped rings of cartilage

outer inter-costal muscles

inner inter-costal muscles

ribs

abdominal cavity

## 4.2 BREATHING MOVEMENTS

### 4.2.1 INSPIRATION or BREATHING IN

Inspiration is an active process. The external intercostal muscles contract, causing the ribs and sternum to move upwards and outwards. At the same time, the diaphragm muscles contract, causing the central part of the diaphragm to move down, so that the diaphragm flattens. These movements of the ribs, sternum, and diaphragm cause the thorax and the lungs to increase in volume, and the air pressure within the lungs to decrease below the external air pressure. As a result the external air pressure forces air from the outside into the lungs.

**Figure 4.2:** *Side view of thorax to show breathing movements.*

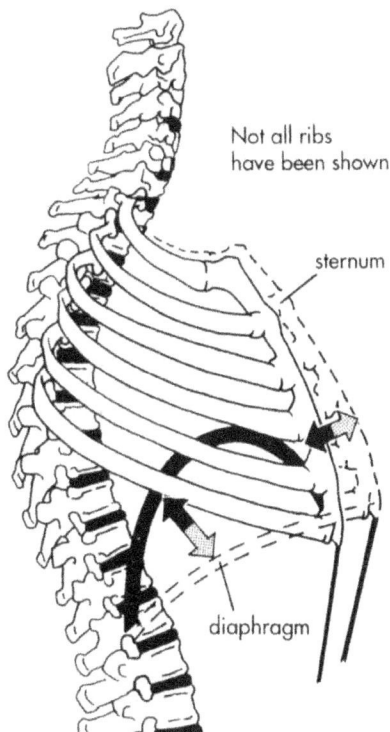

Not all ribs have been shown

sternum

diaphragm

**Muscles relaxed**

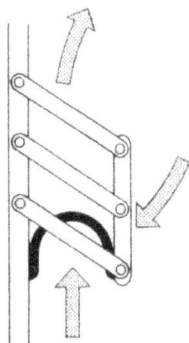

Volume of thorax decreases, pressure increases, air forced out.

Ribs swing down and in.

Diaphragm returns to relaxed position.

**Muscles contracted**

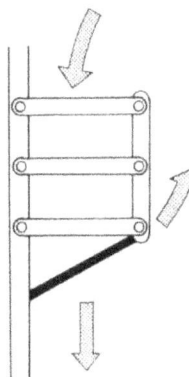

Volume of thorax increases, pressure decreases, external air pressure forces air in.

Ribs swing up and out.

Diaphragm contracts and flattens.

## 4.2.2 EXPIRATION or BREATHING OUT

In normal quiet breathing, expiration is passive. The external intercostal muscles relax, and the ribs and sternum move downwards and inwards regaining their original positions. At the same time, the diaphragm muscles relax, and the central part of the diaphragm moves up, so that the diaphragm regains its domed shape. These movements of the ribs, sternum, and diaphragm, aided by the elastic recoil of the thoracic tissues and abdominal muscles, cause the thorax and the lungs to decrease in volume, and the air pressure within the lungs to increase above the external air pressure. As a result the greater internal pressure forces air out of the lungs.

In forced breathing, for example during exercise, expiration is active. The internal intercostal muscles contract, moving the ribs and sternum downwards more forcibly. The abdominal muscles also contract, increasing the pressure in the abdominal cavity, which helps to force the central part of the diaphragm up.

---

### ❖ *Training and Performance Applications*

*The strength and efficiency of the respiratory muscles are increased as a result of the work they perform in strenuous exercise. Although this is a contributory factor to the 'easing' of breathing difficulties that is seen with increasing fitness, rather surprisingly it is not the major one (the major one being the increasing efficiency of the circulatory system). In heavy exercise the muscles involved in breathing movements can themselves utilise up to 10% of the oxygen uptake. Cramp of these muscles may be one cause of the 'stitch', but this is far from certain in all cases.*

---

## 4.2.3 CONTROL OF BREATHING MOVEMENTS

Breathing movements are under nervous control, which in turn is influenced by chemical and mechanical changes associated with exercise.

Normal breathing movements are controlled by involuntary, automatic, rhythmic discharges of nerve impulses from the regions of the brain known as the **respiratory control centres**.

The **inspiratory control centre** sends motor nerve impulses to the external intercostal muscles of the ribs, and the muscles of the diaphragm, causing them to contract, and inspiration to occur. This active phase of the inspiratory control centre lasts for about two seconds. After this it automatically stops sending motor impulses for about three seconds, during which time the muscles of inspiration relax and expiration follows passively.

The basic rhythm of nervous control by the respiratory control centres is influenced by sensory inputs so that breathing alters in response to the demands of the body during exercise.

During forced breathing, the expiratory control centre is activated, stimulating the internal intercostal muscles and abdominal muscles to contract, resulting in forced expiration.

Ventilation of the lungs is a function of the rate and depth of breathing. Early on in exercise, ventilation is mainly increased by an increase in the depth of breathing, whilst later it is the rate of breathing that increases more.

Any increase in the acidity of the blood *(decrease in blood pH)* causes an increase in ventilation. A decrease in pH results mainly from an increase in the production of lactic acid and carbon dioxide *(an 'acid' gas)*, due to the increase in the rate of respiration associated with exercise. The level of acidity is monitored by chemoreceptors in the main aorta and in the carotid bodies in the arteries of the neck. Indeed, the sensitivity of the carotid bodies to a decrease in pH is thought to be a major factor controlling the increase in ventilation rate, during exercise, along with nervous stimulation of the respiratory control centres from the contracting muscles and moving joints.

A degree of voluntary control can be exerted over breathing movements, which is necessary for speech. However, once the pH decreases below, and the carbon dioxide in the blood rises above, a certain level, the inspiratory control centre is stimulated, and voluntary control is over-ridden. It is these factors rather than shortage of oxygen that act as the main stimulus for ventilation, and limit breath holding activities.

The blood has a certain buffering capacity, ie under normal conditions it can resist changes in its pH *(acidity or alkalinity)*, restricting the changes in blood pH to between 7.2 and 7.45. As the lactate accumulates in the blood, it is buffered (initially at least) by combination with the hydrogen carbonate ions present in the plasma, increasing the amount of carbon dioxide in the blood.

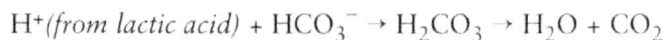

$$H^+ \text{(from lactic acid)} + HCO_3^- \rightarrow H_2CO_3 \rightarrow H_2O + CO_2$$

An increase in the level of carbon dioxide in the blood stimulates the chemoreceptors in the carotid bodies, but its main effect is by direct stimulation of the respiratory control centres in the brain, resulting in an increase in the ventilation rate, and therefore more carbon dioxide being exhaled by the lungs.

Past a certain point the buffering capacity of the blood *(due also to the plasma proteins and the haemoglobin in the red blood corpuscles)* is saturated, and in strenuous exercise the pH within the muscles can drop to 6.4, and the arterial blood pH to about 7.0 or just below, as a result of the accumulation of lactic acid.

Ventilatory Equivalent of $O_2$ 25

$\dot{V}_E/\dot{V}_{O_2}$ Rate of Breathing Over Rate of $O_2$ used in $dm^3.min^{-1}$

These curves reflect the fact that $CO_2$ production and an associated drop in blood pH (increase in acidity) are the main stimuli to increase the ventilation rate.

Ventilatory Equivalent of $CO_2$ 24

$\dot{V}_E/\dot{V}_{CO_2}$ Rate of Breathing Over Rate of $CO_2$ used in $dm^3.min^{-1}$

Graph to illustrate the ventilatory break point or anaerobic threshold.

Ventilation rate increases more than rate of $O_2$ uptake

Rate of $CO_2$ production greater than ventilation rate

Anearobic Threshold or Ventilatory Break Point

**Intensity of Effort** (arbitrary units)

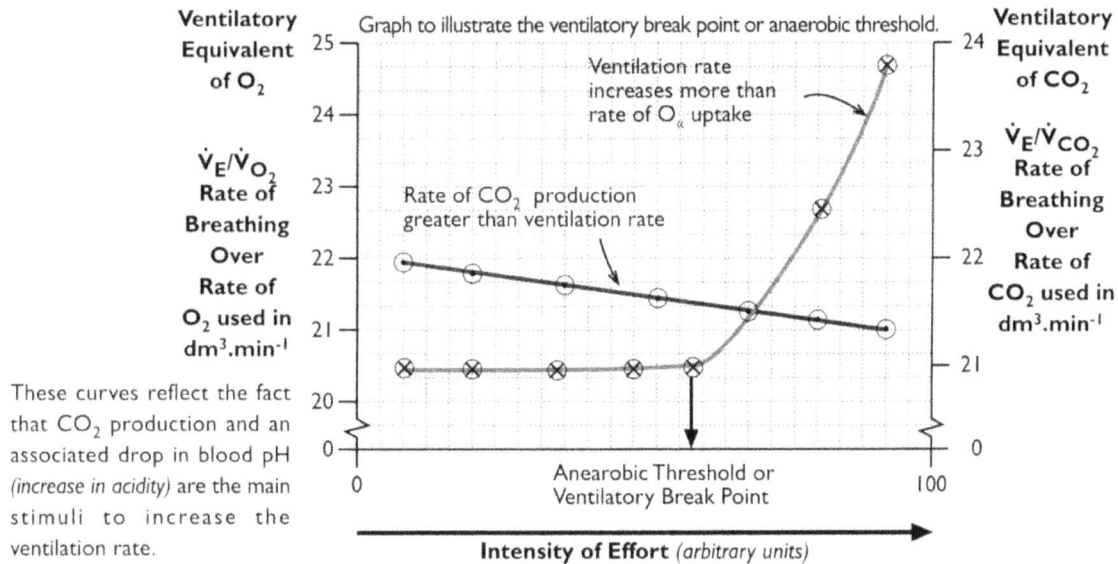

### ❖ Training and Performance Applications

*In terms of aerobic endurance the greater the aerobic capacity and the greater the percentage of the aerobic capacity at which lactate threshold occurs, the better will be the performance. However the exercise intensity just above the lactate threshold is of greater significance to the performance than the aerobic capacity itself. For example if two endurance athletes have equal aerobic capacities, the one who can use the largest percentage of that aerobic capacity without accumulating lactate will have the better performance, all other things being equal. In fact some World ranked endurance athletes have had a relatively low aerobic capacity. Both can be improved by training, but the greatest gains can be made in the percentage aerobic capacity at which lactate threshold occurs. Thus highly trained athletes can perform for long periods at high levels of their aerobic capacity before lactate threshold and exhaustion occur.*

Typically the ventilation of the lungs during exercise increases in proportion to the increase in heart rate and to the uptake of oxygen, however at a certain point the rate of ventilation of the lungs increases more rapidly than the heart rate and the oxygen uptake. This point, at which ventilation 'breaks away' from the heart rate and oxygen uptake, is known as the '**ventilation break point**'. Past this point, the rate of ventilation increases faster than the uptake of oxygen.

The ventilation break point is used in non-invasive estimates of the lactate threshold *(although in this case it is often still referred to as the anaerobic threshold)*. However, the correlation between ventilation and lactate threshold is poor. The difference can be explained by the fact that ventilation is affected by many factors, and not only by lactate. The chemoreceptors in the carotid bodies and the aorta are also sensitive to levels of oxygen in the blood, although less so than to levels of pH and carbon dioxide. A decrease in oxygen below a certain level stimulates an increase in ventilation.

Stretch receptors in the bronchi and the bronchioles are stimulated by the expansion of the lungs. Nerve impulses pass from these along sensory nerves to the inspiratory control centre, causing the reflex inhibition of the inspiratory motor nerve impulses. Therefore the muscles of inspiration relax, and expiration follows passively. However, this mechanism is not thought to be part of the normal control of breathing, but operates to protect the lungs from over-inflation.

Other sensory receptors *(proprioceptors)* in the skeleton and muscles, also feedback sensory information to increase breathing movements during exercise. In conclusion it should be noted that the changes in blood levels of carbon dioxide, pH, and oxygen, together with mechanical feedback, cannot fully account for the large and rapid changes in ventilation experienced during exercise. Other factors increasing ventilation, include a rise in body temperature, and an increase in the blood levels of adrenaline and/or noradrenaline. The emotions are also involved, possibly via the sympathetic nervous system and the adrenal glands. The involvement of these may help explain the anticipatory rise in ventilation that can occur before the start of exercise.

## 4.3 LUNG VOLUMES

Breathing movements result in certain volumes of air passing into and out of the lungs. These volumes vary between individuals, but a healthy, active mature male of average height would have figures approximating to those set out below.

During quiet breathing when at rest, about 500 cm$^3$ of air can be breathed in and out, and this is known as the **tidal volume**. Of this 500 cm$^3$, about 350 cm$^3$ reaches the alveoli where gaseous exchange occurs. The remaining 150 cm$^3$ fills the pharynx, larynx, trachea, bronchi, and bronchioles, which compared to the alveoli are poorly supplied with blood capillaries, so that very little, if any, gaseous exchange occurs here. This space where no gaseous exchange occurs is known as the anatomical dead space, and the air that occupies it as the dead air volume.

The 350 cm$^3$ of the tidal volume that reaches the alveoli, mixes with the air that remains in the alveoli in quiet breathing, the stationary air, which can be up to 2500 cm$^3$. Thus the alveolar air consists of about 350 cm$^3$ of fresh air mixed with up to 2500 cm$^3$ of air that has undergone gaseous exchange with the blood.

Therefore the composition of alveolar air does not fluctuate widely during the breathing cycle, which in turn ensures that gaseous exchange with the blood also remains fairly constant throughout the breathing cycle. However, when the depth of breathing increases during exercise more 'fresh air' does reach the alveoli.

By breathing in as deeply as possible an extra volume of air, in addition to the tidal volume and the stationary air, can be taken into the lungs. This is the **inspiratory reserve volume** and can be up to 2000 cm$^3$.

By breathing out as forcibly as possible after returning to normal breathing, it is possible to expel some of the stationary air in addition to the tidal volume. This is known as the **expiratory reserve volume** and can be up to 1500 cm$^3$. However the lungs cannot be emptied entirely, otherwise they would collapse, and this remaining air is known as the **residual volume** which can be up to 1500 cm$^3$.

These volumes can be used to calculate the various lung capacities.

### 4.3.1 LUNG CAPACITIES

The **inspiratory capacity** *(up to 2500 cm³)* is the sum of the tidal volume and the inspiratory reserve volume.

The vital capacity *(up to 4000 cm³)* is the sum of the **inspiratory reserve volume**, the tidal volume, and the expiratory reserve volume . The vital capacity represents the maximum total amount of air that can be moved into and out of the lungs by one inhalation and one exhalation.

The **total lung capacity** *(about 5500 cm³)* is the sum of all the lung volumes.

These volumes and capacities vary between individuals and are generally smaller in females, due to smaller body size, than in males. In fact up to the age of 25, vital capacity and total lung capacity are more closely correlated with height and body surface area than any other factors.

---

### ❖ *Training and Performance Applications*

*Perhaps surprisingly there is no simple relationship between the vital capacity, maximum oxygen uptake ( $\dot{V}_{O_2}$ max), or aerobic performance. Neither does aerobic training have as much effect on the lungs as might be expected. The lung capacities may be increased a little, and forcible exhalation may increase in power, but compared to other adaptations to training, the effects of training on the breathing system are less significant. The marked relief in the stress of breathing experienced with increasing fitness, is more significantly related to improvements in cardiovascular fitness, than with improvements in breathing mechanisms themselves. The efficiency of the breathing mechanisms is not normally a limiting factor in aerobic exercise. Only in elite athletes might the lungs be the limiting factor for maximum exercise.*

---

These lung volumes and capacities are measured by means of a spirometer, of which there are two main types. The most commonly used is the hinged box type with a fixed pen, in which breathing-in causes the trace to move down, and breathing out causes the trace to move up *(as shown in Figure 4.3, p 86)*. The other type, using a cylinder and pulley system, gives a trace which moves up when breathing in, and down when breathing out, thus swapping the position of the expiratory and inspiratory measures shown in Figure 4.3.

**Figure 4.3:** *Spirometer trace.*

**Vital Capacity Trace**

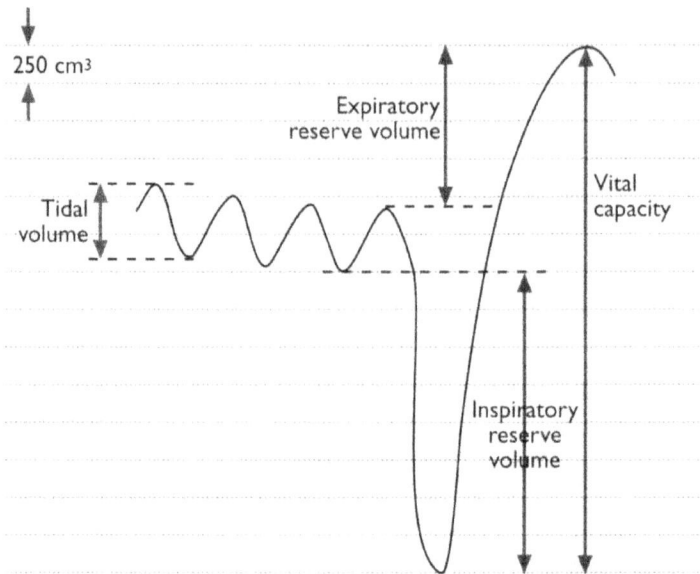

This trace does not show the total lung volume and residual volume, which are not easily measured. With increasing age the residual volume increases from about 20% of the total lung volume to more than 30%.

The volume of air breathed, in one minute, is known as the **Minute Volume** or **Minute Expiratory Ventilation** *(V̇E)*. That is the tidal volume multiplied by the breathing rate per minute.

During quiet breathing this can be:

$$500 \text{ cm}^3 \times 16 = 8000 \text{ cm}^3 \text{ } (8.0 \text{ } dm^3)$$

The tidal volume increases from about 12% of the vital capacity at rest, to about 50-60% of the vital capacity during exercise, and the breathing rate can increase to up to 50 per minute. Therefore the maximum minute volume in a fit male can rise to:

$$\dot{V}_E \text{ max} = 3000 \text{ cm}^3 \times 50 = 150 \text{ } 000 \text{ cm}^3$$

The **Maximum Voluntary Ventilation** *(MVV)* is the volume of air that can be breathed in 15 seconds of forced rapid and deep breathing, multiplied by 4, to give a figure for one minute. In males this can typically be up to 200 000 cm$^3$ per minute, which is a greater volume than can be breathed during maximum exercise.

The **Timed Vital Capacity** or **Forced Expiratory Volume** *(FEV$_1$)* is that volume of air in dm$^3$ *(litres)* that can be expired in one second. It is an indicator of the power of the lungs to breath out. The FEV$_1$ is also expressed as a percentage of the forced vital capacity or *(FEV$_1$/FVC ratio)* and gives a measure of the overall resistance to air flow *(normal value 85%, below 65% in patients with pulmonary disorders)*. Resistance to air flow in the air tubes accounts for up to 80% of the total resistance of the lungs to ventilation, particularly during expiration when the airways tend to collapse. The remainder of the resistance is due to the elasticity or **compliance** of the lung tissues.

❖ *Training and Performance Applications*

*During exercise, adrenaline and an increase in sympathetic nervous activity, cause the smooth muscle fibres in the walls of the airways to relax, resulting in dilation of the airways and a decrease in the resistance to air flow. For this reason restrictions exist over the use of asthmatic inhalers by competitors, as they work on the principle of dilating the airways. In cold air, the airways and pulmonary capillaries constrict, decreasing the efficiency of ventilation and gas exchange.*

## 4.4 GAS EXCHANGE AT THE LUNGS

Gas exchange occurs between the air in the alveoli of the lungs and the blood in the capillaries present in the walls of the alveoli of the lungs.

### 4.4.1 DIFFUSION

A substance, as a gas or in solution, will move by diffusion from a region of its higher concentration to a region of its lower concentration, down a concentration gradient, until an equilibrium is reached. An increase in temperature increases the speed of diffusion. An increase in the differences of concentration between two regions, that is an increase in the concentration gradient, increases the amount of substance diffusing, but not its speed of diffusion.

### 4.4.2 PARTIAL PRESSURE

The continuous random movement of particles of a gas means that the particles will occasionally collide with each other and with the walls of any structure enclosing them within a space, thus exerting a pressure. The number of such collisions, and therefore the pressure, is proportional to the amount of substance present. In mixtures of gases, e.g. air, each substance exerts a partial pressure in that mixture proportional to the amount of that substance in the mixture.

The partial pressure of a gas is a better measure of the amount of gas present than quoting its concentration as a percentage *(e.g. 21% oxygen in air)*, because, for example, although the *percentage* of oxygen remains the same in air under different conditions, the *amount* of oxygen does not. At atmospheric pressures less than at sea level e.g. at altitude, there is still 21% oxygen in air, but as the air is less dense there is a smaller amount of oxygen.

The partial pressure of gases can be calculated as follows:

$$\text{partial pressure of gas (p)} = \text{barometric pressure of gas} \times \text{fractional concentration}$$

*(The units of pressure are either millimetres of mercury (mm Hg) or kilo Pascals (kPa) ).*

The normal barometric pressure of dry atmospheric air at sea level is 760 mm Hg *(101.3 kPa)*.

So at normal atmospheric pressure at sea level the $pO_2$:

$$pO_2 = 760 \times \frac{21}{100} = 159.6 \text{ mm Hg} \ (21.273 \text{ kPa});$$

and the partial pressure of carbon dioxide is:

$$pCO_2 = 760 \times \frac{0.04}{100} = 0.3 \text{ mm Hg}$$

### 4.4.3 EXCHANGES IN THE ALVEOLI

Gaseous exchange occurs by diffusion between the alveolar air and the blood in the capillaries in the walls of the alveoli, until equilibrium is reached.

---

❖ *Training and Performance Applications*

*Strenuous activity can result in an increase in the size of the alveoli, and an increase in the blood supply in the capillary beds of their walls. Both of these developments increase the surface area for gaseous exchange by diffusion between the alveolar air and the blood, and thus increase the efficiency of breathing. However as has been mentioned earlier (Section 4.3.1, p 85), this is rarely a limiting factor in exercise.*

---

The concept of partial pressures also applies to the diffusion of gases from a gas mixture to gases in solution, and vice versa. Gases in contact with a liquid dissolve into solution by diffusion until an equilibrium is reached between the gases and the gases in solution. At equilibrium the partial pressure of the gases is equal in both gaseous and liquid phases, and gases are diffusing into and out of the two phases at an equal rate. These considerations apply to the exchange of gases between the alveolar air and the blood circulating in the capillaries of the alveoli walls.

Blood entering the capillaries of the alveoli from the pulmonary arteries has a lower oxygen and a higher carbon dioxide content, than the air in the alveoli. Therefore carbon dioxide diffuses out of the blood into the alveoli, and oxygen diffuses into the blood from the alveoli, down their partial pressure gradients.

The oxygen diffusing into the blood dissolves in and diffuses through the surface film of moisture on the surface of the alveoli, through the thin walls of the alveoli, through the thin wall of the capillaries, through the plasma, and through the red blood corpuscle membrane to combine with the haemoglobin to form **oxyhaemoglobin**.

About 250 cm³ of oxygen per minute diffuses in over the lungs at rest, and this can be increased up to about twenty times this amount during exercise.

The carbon dioxide diffuses in the opposite direction to the oxygen, from the blood plasma into the alveoli.

Carbon dioxide molecules are larger than those of oxygen and therefore diffuse more slowly in the gas phase, but carbon dioxide is much more soluble than oxygen and therefore diffuses about twenty times more readily than oxygen across the capillary-alveolar barrier. This explains why, although the pressure gradient of carbon dioxide is only 6 mm Hg compared to 60 mm Hg for oxygen, sufficient exchange of $CO_2$ still occurs.

**Figure 4.4:** *Alveoli and gas exchange.*

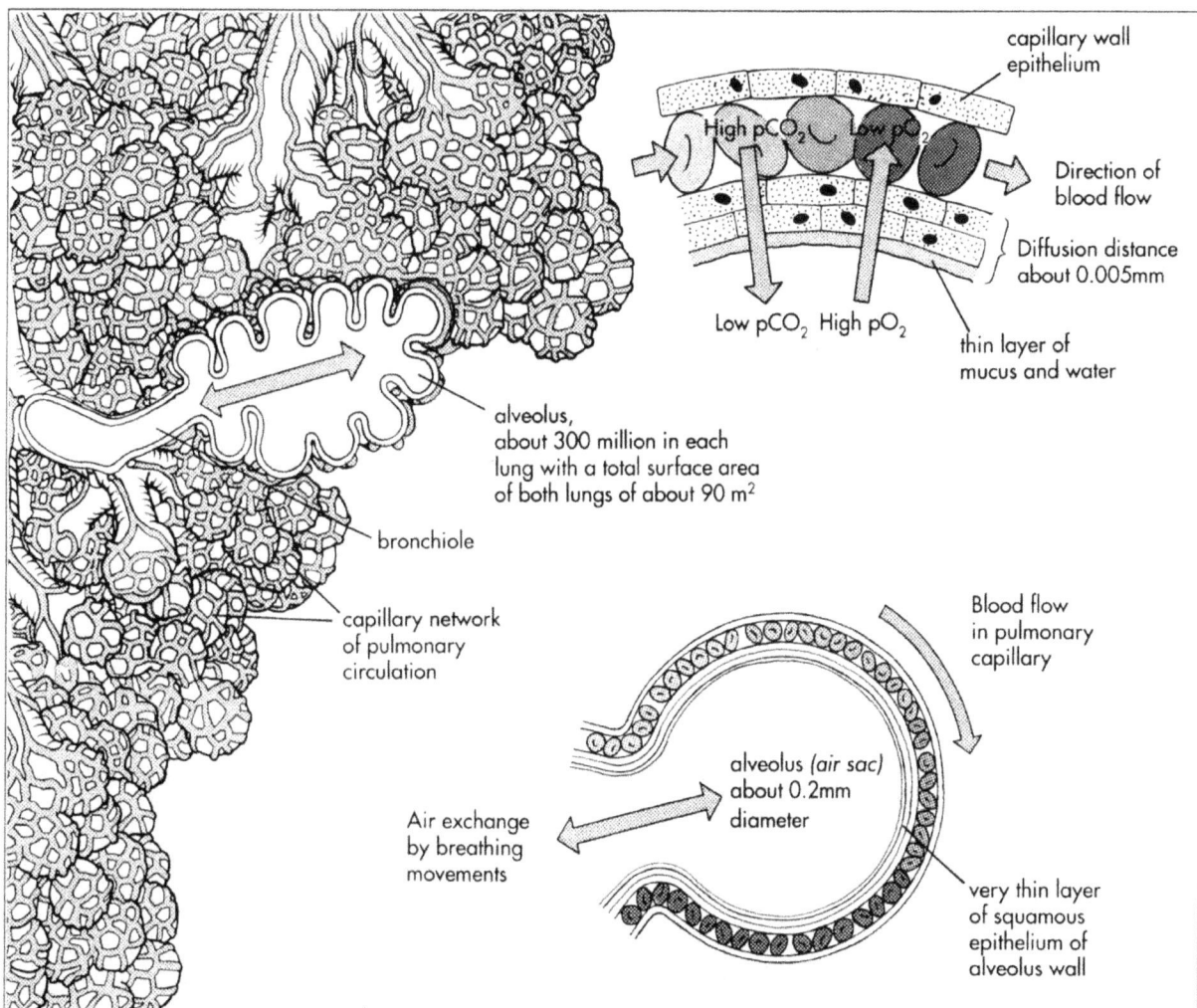

**Table 4.1:** *Exchanges between alveolar air and pulmonary blood at rest.*

|  | Blood in pulmonary artery | | Alveolar air | Blood in pulmonary vein | |
|  | $cm^3$ per 100 $cm^3$ blood | mm Hg | mm Hg | $cm^3$ per 100 $cm^3$ blood | mm Hg |
|---|---|---|---|---|---|
| $O_2$ | 12.5 | 40.0 | 98.0 | 19.0 | 96.0 |
| $CO_2$ | 56.0 | 46.0 | 40.0 | 50.0 | 40.0 |

**Table 4.2:** *Differences between inspired and expired air.*

|  | Dry Inspired Air | | Trachea | Alveolar air | | Expired air | |
|  | % | mm Hg | mm Hg | % | mm Hg | % | mm Hg |
|---|---|---|---|---|---|---|---|
| $O_2$ | 21.0 | 159.6 | 149.0 | 13.8 | 98.0 | 16.4 | 116.2 |
| $CO_2$ | 0.04 | 0.3 | 0.3 | 5.5 | 40.0 | 4.0 | 28.5 |
| $N_2$ | 79.0 | 600.0 | 564.0 | 80.7 | 575 | 79.6 | 567.5 |

Expired air is a mixture of the alveolar air and the air that occupied the dead space. The inspired air becomes saturated with water vapour, and this has the effect of decreasing the partial pressure of oxygen or $pO_2$ by about 10 mm Hg. The $pCO_2$ is less affected as there is such a small amount, and the $pN_2$ is more affected as there is a larger amount. Nitrogen plays no part in metabolism, but its percentage increases *(by +1.7% in alveolar air)* in proportion to the decrease in volume caused by more oxygen being absorbed *(7.2%)* than carbon dioxide being given out *(5.5%)*, a difference of -1.7%. More oxygen is absorbed than carbon dioxide is given out as a result of the oxidation of fats and proteins, which consume more oxygen than they give out carbon dioxide.

### 4.4.4 FACTORS AFFECTING GAS EXCHANGE IN THE ALVEOLI

The exchange of gases by diffusion across the alveolar-capillary barrier depends on many factors.

Although the most important factor is the difference in partial pressure of gases across the barrier, the surface area and thickness of the barrier are also significant. The 300 million alveoli of the lungs have a huge surface area of about 70 $m^2$, and their walls are made up of very thin flat lining cells *(epithelium)*. The capillaries of the pulmonary circulation are sandwiched between the thin walls of adjacent alveoli, causing the alveoli surfaces to be ridged and corrugated, which further increases their surface area. These capillary networks are the most dense in the body. The capillaries also have very thin walls, so that the total thickness of cells between the air and blood can be as little as 0.001 mm.

At rest not all parts of the lung are equally ventilated, nor adequately supplied with blood. The upper parts of the lungs in particular, are generally undersupplied with blood. Such parts of the lung make up the **physiological dead space**, as no significant gas exchange occurs. During exercise, there is an increase in blood supply to the lungs, so that all regions are well supplied.

At rest, the speed of circulation of the blood through the lungs is such that the contact time between the blood and the wall of an alveolus is about 0.75 seconds. During exercise, although this is reduced to about 0.35 seconds, the rate of gaseous exchange by diffusion is such that the blood still comes into equilibrium with the air in the alveoli.

The absorption and utilisation of oxygen from the blood by the tissues, especially the working muscles, leads to a difference in the oxygen content of arterial and venous blood. This is referred to as the **arterio-venous difference** or **a-$\bar{v}O_2$ diff per 100 cm$^3$ blood**.

*(The $\bar{v}O_2$ refers to the oxygen content of mixed venous blood, that is blood returning from both active muscles and other less active tissues.)*

At rest most blood is returning from the gut, its associated organs, and the kidneys, which extract relatively little oxygen from the blood. However, during exercise blood is shunted from these organs to the active muscles, which do extract large amounts of oxygen, so that the oxygen content of the mixed venous blood decreases towards that of the blood leaving the active muscles, increasing the a-$\bar{v}O_2$ diff. The a-$\bar{v}O_2$ diff is larger in trained individuals as a result of the adaptations to greater oxygen usage in the muscles. This increase in the a-$\bar{v}O_2$ difference increases gaseous exchange by diffusion in the alveoli, because venous blood coming from the body via the heart and the pulmonary arteries to the lungs contains more carbon dioxide and less oxygen than at rest, and this increases the diffusion gradients between the blood and the air in the alveoli.

During exercise, oxygen uptake can increase up to 5000 cm$^3$ per minute, as a result of the increase in ventilation of the lungs, the increased blood flow through the pulmonary circulation, and increased rates of diffusion.

## 4.5 TRANSPORT OF OXYGEN

The oxygen absorbed into the blood in the capillaries in the walls of the alveoli of the lungs combines with haemoglobin in the red blood corpuscles to form oxyhaemoglobin. The percentage saturation of haemoglobin with oxygen is proportional to the $pO_2$. At rest and even during maximum exercise, at sea level, the haemoglobin in the blood leaving the lungs in healthy subjects is virtually saturated with oxygen. Therefore the capacity and functioning of the lungs is not normally the limiting factor in exercise.

❖ *Training and Performance Applications*

*As even during maximum exercise the haemoglobin in the blood leaving the lungs is still virtually saturated with oxygen, there is no significant advantage to be gained by breathing pure oxygen at sea level. If pure oxygen is breathed under high pressure, so much oxygen may dissolve in the plasma that the needs of the tissues may be supplied in this way, and the oxyhaemoglobin does not give up its oxygen. This in turn interferes with the transport of carbon dioxide, and breathing oxygen under these conditions for long periods is harmful. Under normal conditions, little oxygen dissolves in the plasma, but what does is important, as a decrease in this dissolved oxygen stimulates an increase in ventilation, and over a longer period, an increase in red blood cell production.*

The concentration of red blood cells and their haemoglobin affects the amount of oxygen taken up by the blood. The red blood cells typically make up about 40-45% of the blood volume. These concentrations increase during exercise, as more fluid moves from the plasma to the tissues, and more water is lost from the plasma as sweat. Also long term endurance training results in some increase in red blood cells and hence haemoglobin. (*Although as mentioned earlier the greater increase in plasma volume can result in an overall decrease in red blood cell count and thus haemoglobin concentration.*) This increase in red blood cells is stimulated by an increased production of **erythropoietin (EPO)**, which is a hormone secreted by the kidneys, which stimulates the red bone marrow to produce more red blood cells.

❖ *Training and Performance Applications*

*EPO can be synthesised artificially, and is readily available. Its use by endurance athletes is suspected. Although its use is officially banned, detection is not yet possible, so the ban is ineffective. Another banned technique for increasing the number of red blood cells is **blood doping**. In blood doping, a volume of blood (about a l dm³ (litre)) is withdrawn from the athlete and stored for about a* month. When the body has replaced the lost blood, the extracted red blood cells are re-infused prior to the event, thus increasing the oxygen carrying capacity of the blood. However, a substance 2,3-DPG (levels of which are increased with training) in red blood cells , which increases the unloading of oxygen from oxyhaemoglobin to the tissues, deteriorates in stored blood, and the re-infused blood may not be as effective as supposed. ('Phosphate loading' by taking in sodium phosphate has been claimed to increase levels of 2,3-DPG, increasing aerobic capacity and lowering blood lactate levels.) Problems may also arise from the increased viscosity of the blood (due to the greater number of red blood cells) causing difficulties in circulation, such as raised blood pressure and decreased cardiac output. There is conflicting evidence as to the effectiveness of blood doping, and whether the abuse is widespread or not is not known, as once again detection is not yet possible. (A classic case, however, occurred when a well known athlete took steroids in training, but came off them in time to be clear for any testing on the day of the event. However, he was also blood doping, and received the steroids back in the re-infused blood. These were detected as a result of testing after the competition, and he was subsequently banned). **Altitude training** is based on the same principle of increasing the red blood cell count and hence the amount of haemoglobin in a given volume of blood, and therefore increasing the amount of oxygen transported. The red blood cell volume (haematocrit) can increase from 40-45% at sea level to 59% after 6 weeks at 4500 m. This compares to 58% as a result of the use of EPO. The substance 2,3 DPG, muscle capillaries, and myoglobin are also increased. All these changes are typical adaptations to oxygen stress seen in normal aerobic endurance training, but at altitude the oxygen stress is greater. As altitude increases, the $pO_2$ in air decreases as the air becomes less dense, so that at an altitude of about 5500 m, the amount of oxygen in a given volume of air is half that at sea level. There is some compensation in the fact that the thinner air at altitude is easier to breathe, so that the maximum breathing capacity is greater than at sea level, and this may increase the efficiency of the breathing mechanisms by more than could be achieved at sea level. Increased ventilation removes more $CO_2$ from the blood, so the blood becomes more alkaline on arrival at altitude. As the body adjusts to altitude this excess alkalinity is excreted by the kidneys, with the result that on return to sea level the blood has less buffering capacity. About 3-4 weeks of altitude training are considered to be necessary to achieve the necessary gains, and about 7-10 days to reacclimatize at sea level, but there is considerable individual variation in response. The case for altitude training is complex and far from proven for all athletes. The ideal situation would appear to be to live at altitude and train at sea level, where quality of training can be maintained. Performance in the power events such as sprints, jumps, and throws, that are not dependent on maximum oxygen uptake, benefit from the lower air resistance of the less dense air.

Males normally have a higher level of haemoglobin *(14-18 g/100 cm³ blood)*, than females *(12-16 g/100 cm³ blood)*, and along with the larger blood volume in the male, this is an important factor in the greater aerobic power of males.

An increase in temperature, and a decrease in pH *(increase in acidity)* decrease the affinity of haemoglobin for oxygen. These factors are of importance in the release of oxygen from oxyhaemoglobin to the tissues.

**Figure 4.5:** *Association and dissociation curves of haemoglobin. Curve* **1** *operates at the lungs, and curve* **2** *at the tissues. As blood gains $CO_2$ from the tissues the functional curve gradually moves from* **1** *to* **2***, ie in the presence of $CO_2$ more $O_2$ is liberated to the tissues. The curve is also shifted to the right by an increase in temperature and a decrease in pH, both of which occur in active muscles.*

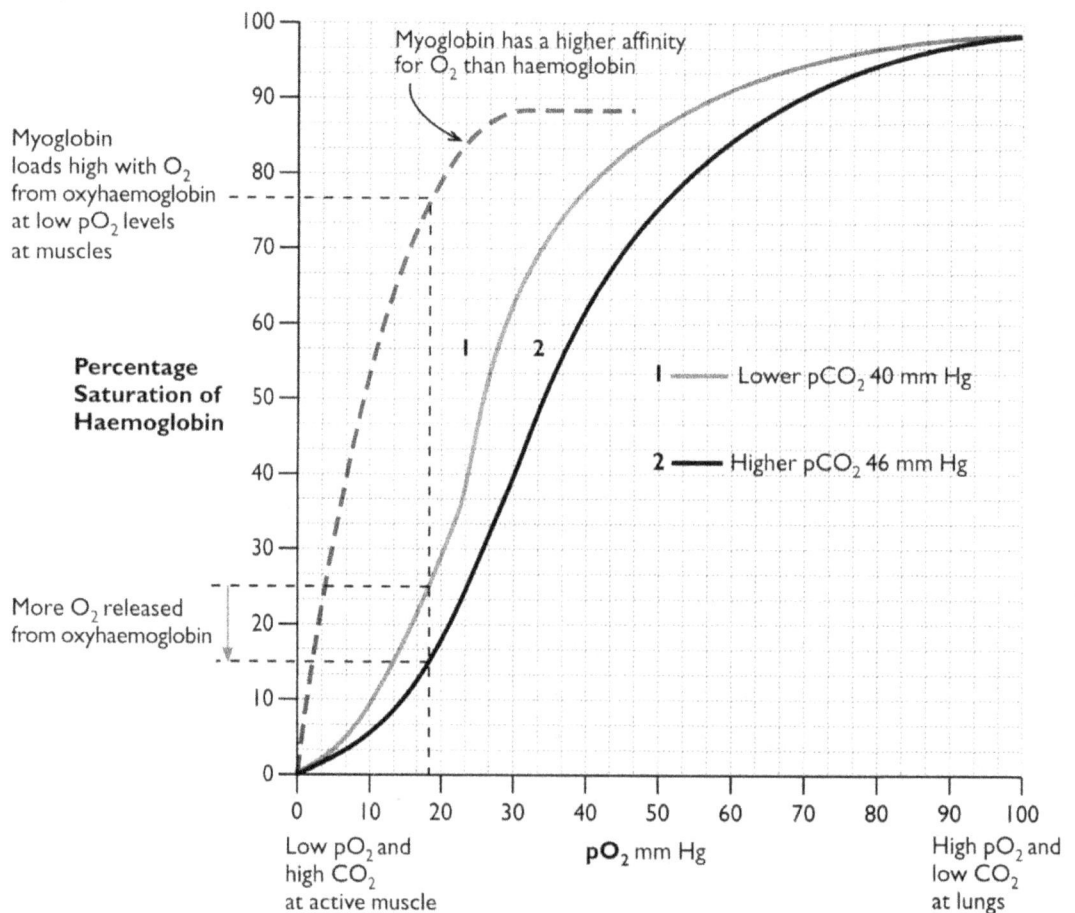

## 4.6 GAS EXCHANGE AT THE TISSUES

As at the lungs, partial pressure gradients determine the diffusion of oxygen and carbon dioxide between the blood and the tissues.

Tissues vary widely in their demands for oxygen and in their production of carbon dioxide, particularly during exercise when the muscles become active.

**Table 4.3:** *Gas exchange at the tissues at rest*

| | Arterial blood | | Tissues | Venous blood | |
|---|---|---|---|---|---|
| | cm³/100 cm³ blood | mm Hg | mm Hg | cm³/100 cm³ blood | mm Hg |
| $O_2$ | 19.0 | 96.0 | 0 - 40 | 12.5 | 40.0 |
| $CO_2$ | 50.0 | 40.0 | 50 - 90 | 56.0 | 46.0 |

Levels of oxygen and carbon dioxide in arterial and venous blood.

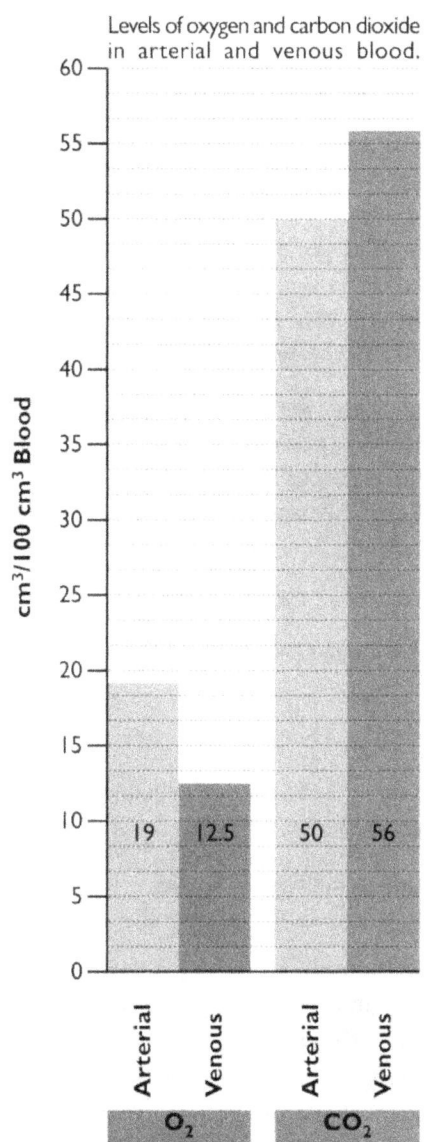

The low oxygen partial pressure in the muscles during exercise increases the diffusion gradient between them and the red blood corpuscles, thus encouraging the dissociation of the oxyhaemoglobin and the release of oxygen in solution to diffuse to the tissues. This release of oxygen from the oxyhaemoglobin is further encouraged during exercise by an increase in temperature, and a decrease in pH as a result of lactic acid and $CO_2$ production, both of which occur in working muscles

The released oxygen diffuses down its diffusion gradient, through the blood corpuscle membrane, through the plasma, through the capillary wall in the tissue fluid, into the muscle fibre cytoplasm and into the mitochondria. The diffusion distance between the blood and the muscle fibres is decreased during exercise as more capillary beds open. Also the movement of tissue fluid to the tissues carries oxygen along in solution from the capillaries.

Muscle fibres, especially the slow twitch types, contain the pigment myoglobin, which has a higher affinity for oxygen than haemoglobin, and increases the uptake of oxygen by the muscle fibres. Myoglobin as oxymyoglobin acts as a store of oxygen within the muscle fibres which can be exploited quickly at the onset of exercise, and when the blood flow in the muscle is reduced by compression of the capillaries as the muscle is contracting.

At rest, only about 25% of the oxygen transported in the blood is used, so that even venous blood is still approximately 75% saturated with oxygen. During maximal exertion up to 85% of the transported oxygen may be used.

### ❖ *Training and Performance Applications*

*Endurance training increases the capillary supply and the amount of myoglobin in the muscles, especially those of the slow twitch type. It also results in an increase in the number, size and enzyme activity of the mitochondria, which increases the potential of the muscles to use oxygen (oxidative potential). The rate of gas exchange at the muscles is an important limiting factor in aerobic exercise.*

*The 'economy' of exercise is measured as the oxygen cost of performing, eg running, at a certain pace. Over a period of training the economy of running can be improved by up to 5%.*

*The oxygen cost of the resting metabolic rate is on average about 3.5 cm³ O₂ . kg⁻¹ . min⁻¹. This is referred to as one metabolic equivalent (MET), and can be used as a unit of oxygen cost of various activities, eg walking at 4 mph = 6.5 METS, and running at 10 mph = 15.0 METS.*

## 4.7 TRANSPORT OF CARBON DIOXIDE

The carbon dioxide, produced as a waste product of aerobic respiration in the mitochondria, diffuses in solution down its concentration gradient, from the high $pCO_2$ in the mitochondria to the lower $pCO_2$ in the blood. The greater the difference in $pCO_2$ the greater the amount of diffusion that occurs. Some carbon dioxide will also be carried away in the tissue fluid which is drained from the tissues in the lymphatic system.

About 5% of the total carbon dioxide carried away by the blood is in simple solution in the plasma. As with the oxygen in solution, this is important, as the $pCO_2$ of the blood contributes to the control of ventilation. Up to another 5% may be combined with water to form carbonic acid.

$$CO_2 + H_2O \rightarrow H_2CO_3$$

This is a weak acid and only dissociates slightly into hydrogen and hydrogencarbonate ions:

$$H_2CO_3 \rightarrow H^+ + HCO_3^-$$

Less than 1% of the carbon dioxide combines with plasma proteins to form carbamino compounds.

Some carbon dioxide is also carried in combination with haemoglobin, as carbaminohaemoglobin in the red blood corpuscles.

The greatest proportion of carbon dioxide however is carried as hydrogen carbonate ions $(HCO_3^-)$ in the plasma. These are formed from carbon dioxide that diffuses into the red blood corpuscles. The hydrogencarbonate ions are then released into the plasma. The reverse process occurs at the lungs, and the carbon dioxide is released for exhalation.

The red blood cells are involved in both oxygen and carbon dioxide exchanges and transport, both of which influence each other.

At the tissues, carbon dioxide diffuses into the red blood cells, and as mentioned previously some combines with the haemoglobin, to be carried as carbaminohaemoglobin.

Most however, forms carbonic acid.

$$CO_2 + H_2O \rightarrow H_2CO_3$$

However, red blood cells contain an enzyme that catalyses this reaction and accelerates the formation of carbonic acid. Oxygen is released from oxyhaemoglobin at the same time, and the deoxygenated haemoglobin acts as a hydrogen acceptor, encouraging the dissociation of carbonic acid into hydrogen ions and hydrogencarbonate ions.

$$H_2CO_3 \rightarrow H^+ + HCO_3^-$$

The hydrogen ions continue to be absorbed by the haemoglobin, and the hydrogencarbonate ions diffuse out into the plasma, in exchange for chloride ions which diffuse into the red blood cells from the plasma, in what is known as the 'chloride shift'.

At the lungs these reactions are reversed, the carbon dioxide is released into solution, and diffuses down its diffusion gradient from the blood into the alveoli of the lungs.

# 4.8 MAXIMUM OXYGEN CONSUMPTION

During extreme exercise, there is a level of work beyond which the oxygen uptake of an individual shows no further increase with additional effort. This point represents the $VO_2$ max or maximum aerobic capacity of that individual.

The maximum oxygen consumption or uptake is the greatest rate of oxygen uptake an individual can achieve during exercise, while breathing air at sea level. It is measured as the greatest difference between the amount of oxygen which enters and leaves the lungs. It is expressed as an absolute value, i.e. $dm^3$ per minute, or in relation to body mass, ie in $cm^3$ per kilogram of body mass per minute. For example if an athlete can consume an absolute level of 3.5 $dm^3$ per minute and body mass is 70 kg, the $VO_2$ max can be expressed as:

$$\left( \frac{3.5}{70} \times 1000 \right) = 50 \text{ cm}^3 \text{ per kg per minute}$$

There is a convention whereby a dot over the V represents 'per minute':

$$\dot{V}O_2 \text{ max in cm}^3 \text{ per kg}$$

An example of oxygen uptake in relation to work output

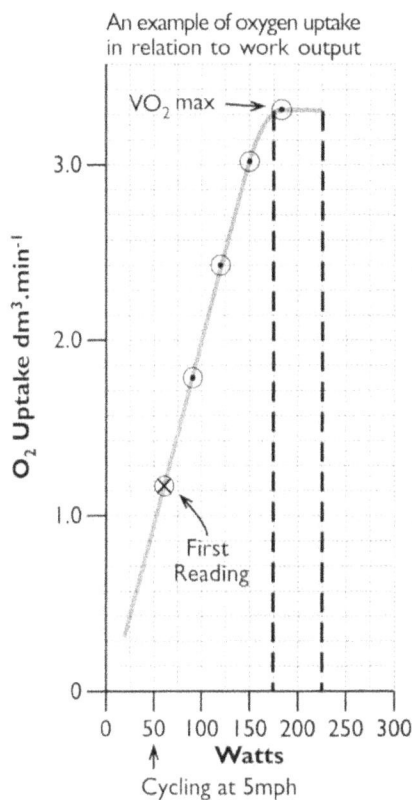

2 minutes at 60 watts and then increments of 30 watts every 2 minutes.
A watt is the unit of rate of work or power.
$1w = 1J.s^{-1}$
A reasonably fit person could cycle about 200 watts for an hour.
Tour de France riders maintain about 250 watts for seven hours.
Peak power outputs are about 1000 watts.

**Figure 4.6:** *Maximal oxygen consumption of male and female elite athletes, and healthy but inactive subjects. (After Saltin B, and Astrand P.O., 1967)*

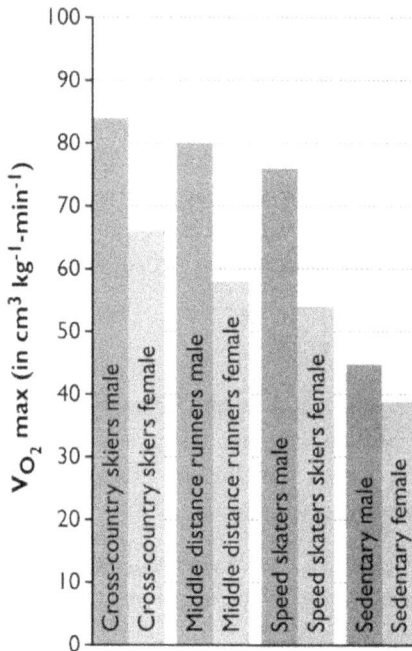

$\dot{V}O_2$ max is commonly expressed in relation to the body mass because most sporting activities are weight bearing in nature. However, in these cases a distinction is not easily, nor often, made between lean body mass *(muscle)* where most of the oxygen is used, and fat mass, where little oxygen is used. Therefore fluctuations in body mass and body composition make the interpretation of training induced changes in $\dot{V}O_2$ max difficult. It is certainly clear that in endurance events such as running and cycling in hilly and mountainous terrain, that great gains in performance can be achieved by minimising body fat.

Where the activity is not weight bearing and the body mass is less important, such as swimming, rowing and flat terrain cycling, $\dot{V}O_2$ max is expressed in absolute terms.

The $\dot{V}O_2$ max is determined by the capacity of the respiratory system to oxygenate blood, the maximum cardiac output, the ability to deliver oxygen to the tissues, and the capacity of the muscles to extract and utilise the oxygen from the blood.

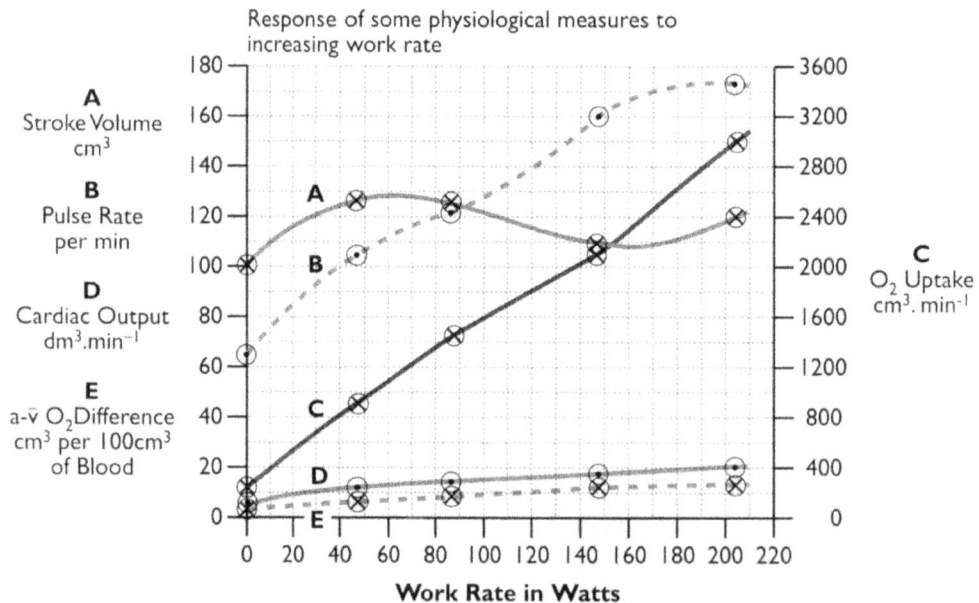

### ❖ Training and Performance Applications

The $\dot{V}O_2$ max represents the maximum ability to absorb, deliver and use oxygen. However, on its own it is not a reliable indicator of performance. Endurance performers of similar ability can have large differences in $\dot{V}O_2$ max, and performance can continue to improve even after $\dot{V}O_2$ max shows no further increase as a result of training. It is the percentage of the $\dot{V}O_2$ max, at which exercise can be sustained over long periods, that is more important. The $\dot{V}O_2$ max gives an idea of the potential aerobic power of a person, while the 'lactate threshold' or OBLA (that level of effort at which lactate begins to accumulate to a significant level in the blood) is more an indication of the current state of fitness as a result of training. The $\dot{V}O_2$ max is largely genetically determined, but can be improved by an estimated 10 – 20% as a result of training. Although all types of endurance training has a general effect on the cardiovascular system, there are effects specific to the muscles being exercised, so that a proportion of the increase in $\dot{V}O_2$ max is due to adaptations in specific muscles. Therefore the type of exercise used whilst investigating the $\dot{V}O_2$ max has an influence on the outcome. For example swimming training has less effect on the $\dot{V}O_2$ max if it is measured whilst running on a treadmill rather than when swimming, and vice versa.

### Major factors determining maximal oxygen uptake ($\dot{V}O_2$ max)

1 **Capacity of respiratory system to oxygenate blood**
   *a)* Efficiency of ventilation, *b)* Lung capacities, *c)* Diffusion capacity

2 **Cardiac Output**
   *a)* Heart Rate, *b)* Stroke Volume, *c)* Cardiac Reserve

3 **Capacity of circulation**
   *a)* Plasma volume
   *b)* Red blood cell number/ haemoglobin concentration
   *c)* Capillary supply to muscles

4 **Muscle metabolism**
   a) Myoglobin concentration
   b) Mitochondria size and number
   c) Amount and activity of respiratory enzymes
   d) Recruitment of efficient ST & FOG muscle fibres.

## SUMMARY

❖ At rest, breathing in is an active process, and breathing out is passive. During exercise and when speaking, breathing out is also an active process.

❖ Normal breathing movements are controlled by involuntary, automatic rhythmic discharges of nerve impulses from respiratory control centres in the brain.

❖ Ventilation of the lungs is a function of the rate and depth of breathing.

❖ The vital capacity is the total volume of air that can be moved over the lungs by the deepest inhalation and the deepest exhalation, that is the sum of the inspiratory reserve volume, the tidal volume, and the expiratory reserve volume.

❖ The blood has a buffering capacity which enables it to resist changes in its pH.

❖ The sensitivity of the carotid bodies to a decrease in pH is a major factor controlling the increase in ventilation during exercise. Carbon dioxide exerts its main effect directly on the brain.

❖ The composition of alveolar air remains relatively constant throughout the breathing cycle, irrespective of whether one is breathing in or out, thus maintaining constant diffusion gradients between the alveolar air and the blood in the capillaries of the lungs.

❖ Gases diffuse down their diffusion gradients as a result of the random movement of their particles, until equilibrium is reached.

❖ Amounts of gas are best measured in terms of their partial pressures, which are exerted as a result of the same forces that cause diffusion, and are directly proportional to the number of particles present.

❖ The alveoli of the lungs, and the capillaries of the pulmonary circulation, represent a huge surface area for gas exchange by diffusion.

❖ The diffusing capacity and ventilatory efficiency of the lungs are improved by training. Therefore the rate and depth of breathing is lower at rest, and increases over a wider range during exercise, than in an untrained subject. However breathing is not normally the limiting factor in exercise.

❖ Although the breathing system is not normally the limiting factor in exercise, a large vital capacity with complete perfusion *(supply)* of blood to all the alveoli, complements a large cardiac output.

❖ A decrease in the oxygen content of mixed venous blood, as occurs in exercise, increases the diffusion gradients in the alveoli, and therefore increases the uptake of oxygen.

❖ Oxygen is mainly carried in the red blood cells in combination with haemoglobin as oxyhaemoglobin. Some is also carried in solution in the plasma.

❖ Carbon dioxide is carried in the blood mainly as hydrogen carbonate ions in the plasma. Some is also carried in combination with haemoglobin in the red blood cells as carbaminohaemoglobin, and in simple solution in the plasma.

❖ $\dot{V}_{O_2}$ max in cm³ per kg per minute *(cm³.kg⁻¹.min⁻¹)*, is a measure of the maximum oxygen uptake an individual can achieve during exercise.

❖ The percentage of the $\dot{V}_{O_2}$ max at which LT occurs is of greater significance to endurance performance than the $\dot{V}_{O_2}$ max by itself.

# 5 EXERCISE, PHYSICAL FITNESS AND HEALTH

## OBJECTIVES

To enable the reader to understand the following:

❖ The definitions of, and relationships between, physical fitness and health.

❖ The components of physical fitness.

❖ The differences in fitness measures between males and females, and their relevance to sport performance.

❖ The effects of the ageing process on fitness measures, and the special considerations relating to children in sport.

❖ The effects of exercise on health.

## 5.1 INTRODUCTION

There are no simple, clear definitions of either of the terms 'fitness' and 'health'. Indeed, such are the difficulties with the definition of 'fitness', that some suggest the term be abandoned.

However, physical 'fitness' can be thought of as the capacity to perform repeated activity with relative success and enjoyment. In terms of sport, sometimes the exertion is so great that it could hardly be considered enjoyable at the time; however in these cases the enjoyment derives from the sense of achievement that follows after the exertion. Any discomfort experienced is generally soon forgotten as a result of the body's rapid recovery. Indeed a common measure of fitness is the speed of recovery after exertion. In less physically demanding skill based sports, physical fitness can be considered as that state which prevents fatigue from interfering with the performance of the skill. Perhaps the clearest definition of physical fitness would be to relate it to the ability of the body to respond to the physical demands made upon it by improving its capacities, that is a person responds to physical demand by becoming "fitter". This is the main principle of training practices, namely progressive overload with adequate recovery leading to adaptation. A battery of tests exists for investigating various anatomical and physiological systems, and the results of these tests can in turn be used to define various levels and categories of fitness *(see accompanying Notes on the Measurement and Testing of Physical Performance)*.

Health is as difficult to define as fitness, but is generally accepted as that state of the body *(which includes the mind)* free of disease, that is compatible with an active and enjoyable life. A sportsperson who drops dead without warning during physical activity may well have been fit, but could be considered as having been unhealthy as they died prematurely. Similarly someone in the early symptomless stages of a fatal disease may well be extremely fit.

Nevertheless, although it is not possible to define the two terms completely satisfactorily, they are dealt with separately here for reasons of clarity of presentation.

## 5.2 PHYSICAL FITNESS

**Table 5.1:** *Components of physical fitness (modified after Eurofit 1988)*

Components or dimensions of physical fitness relate both to sport performance and general health.

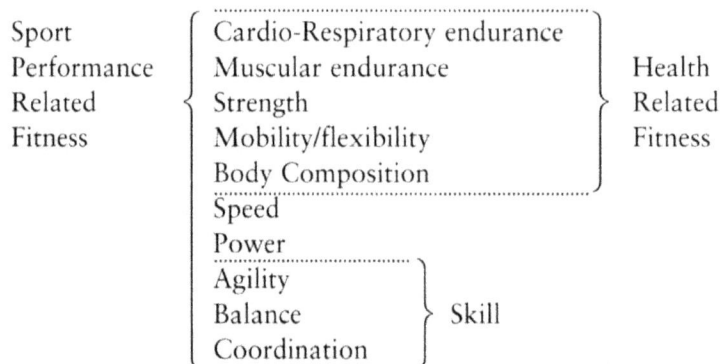

Sport Performance Related Fitness
{
Cardio-Respiratory endurance
Muscular endurance
Strength
Mobility/flexibility
Body Composition
} Health Related Fitness

Speed
Power

Agility
Balance
Coordination
} Skill

### 5.2.1 COMPONENTS OF PHYSICAL FITNESS

(i)   Cardio-Respiratory endurance - the capacity/ability to sustain aerobic work.

(ii)  Muscular endurance - the ability to sustain local muscle action.

(iii) Strength - that force exerted by muscles in a single maximal contraction.

(iv) Mobility/Flexibility - the range of movement about the joints.

(v)  Body composition - the relative amount of body fat compared to lean body mass.

(vi) Speed - the quickness of action of either whole body or part body movements.

(vii) Power - the work performed in a given time, a combination of strength and speed.

(viii) Agility the ability to alter the position of the body and/or its parts, quickly, and accurately in a controlled manner.

(ix) Balance - the ability to control the position of the centre of gravity/mass and maintain it within and above the area of the base of support.

(x)  Co-ordination - the ability to combine the action of various parts of the body in a controlled manner.

## 5.3 DIFFERENCES IN FITNESS MEASURES BETWEEN MALES AND FEMALES

For historical reasons, and not for any reasons of present gender bias, most of the information and data quoted in the text *(unless otherwise stated)* refers to the average male sportsperson.

Females have some essential structural and functional *(physiological)* differences from males with regard to sport performance, and these go some way to explaining their different levels of achievement. There are some who believe that the narrowing gap between the performances of males and females will eventually be closed as a result of the changing roles and continuing emancipation of females in society, especially in developing countries. However, the differences described here would tend to suggest that the gap, at least when comparing top level performances, may well be unbridgeable.

On average, females have a greater percentage body fat *(25%)* than males *(12.5%)*. Although an advantage in long distance swimming as a result of increased buoyancy and heat insulation, increased body fat is a distinct disadvantage in weight bearing endurance events on land. It increases the amount of work necessary to support and move the body, effectively decreases the $\dot{V}_{O_2}$ max when it is measured per kilogram of body mass, and decreases heat loss over the skin by conduction, convection and radiation. Neither does it seem that females have a significantly greater ability to utilise fats in endurance events as was once suggested.

The female skeleton is generally smaller and lighter, with shorter limbs in relation to height, which may be advantageous in events demanding balance and agility. Whilst not strictly physiology, these skeletal differences have physiological implications.

However, the wider pelvis reduces the direct line of thrust from the legs through to the body, and the thighs slant at an angle giving slightly "knock-knees", which further reduces the mechanical efficiency of the running action. *(It is noticeable that all top level female runners are narrow hipped.)* The shoulders are also narrower and more sloped giving a smaller thorax and therefore smaller lung capacities, and the upper body is weaker in relation to the lower body compared with males.

Females typically have a smaller heart, and therefore a smaller cardiac output and faster resting heart rate. There is however no difference in maximum heart rate. The smaller cardiac output reduces the oxygen delivery capacity to the tissues, which reduces endurance capacity when compared to males. There is also a proportionally smaller blood volume. The typical range of haemoglobin concentration in females is between 12-14 g per 100 cm$^3$ of blood, with red blood corpuscles accounting for about 42% of blood volume; compared to the 14-16 g of haemoglobin per 100 cm$^3$ blood and 47% red blood cell volume in males. These figures represent a significant difference in oxygen delivery capacity to the tissues, and therefore in powers of endurance.

Even though there is no significant difference between the sexes in muscle fibre composition; as a result of having lower testosterone levels in the blood, females do not develop as much muscle mass in response to training as males, also they tend to lose strength gains faster than males.

### ❖ *Training and performance applications*

*In attempting to improve performance athlete's may be tempted to resort to 'drug' abuse of various kinds. The most widely abused group of illegal substances are the anabolic steroids. Anabolic means tissue building by means of protein synthesis, and is seen to its greatest extent in the synthesis of muscle protein, which has great relevance to sport performance. Anabolic steroids are all related in some way to the male hormone testosterone, which determines the male characteristics and largely accounts for the superior level of their sport performance as compared to females. For this reason steroid abuse produces the greatest gains in females, but the male determining effects (androgenic) are also more obvious in females, as can easily be observed. Later developed steroids have reduced androgenic effects whilst retaining the anabolic effects. The abuse of anabolic steroids enables a greater training load to be undertaken, with a more rapid rate of recovery, and with a consequent greater training effect; psychological effects which also relate to sport performance, such as increased aggression, are also reported as 'Roid Rage'. Other major hormones that are abused include; Human Growth Hormone and Human Chorionic Gonadotrophin (hCG) which stimulate the body to increase its own production of anabolic steroids. (hCG is normally produced by the developing embryo and is detected in the urine in the pregnancy test. It is also produced by some tumours, and there has been a case of suspected hCG abuse by an athlete which in fact was the result of the presence of a tumour, the early detection of which led to successful treatment), and Adrenocorticotrophic Hormone which stimulates the body to increase its production of corticosteroid hormones which have similar effects to anabolic steroids.*

## 5.4 DIFFERENCES IN FITNESS MEASURES WITH AGE

### 5.4.1 CHILDREN UP TO PUBERTY

Until the onset of puberty boys and girls are sufficiently similar to each other in various physical measures that there is an argument for considering children as a third gender group, rather than small versions of adult females or males. This is very important when considering 'training' regimes for children, so much so that many would argue that the specialised training of children is not to be recommended. Increasingly children in sport are being treated as miniature adults and exposed to undesirable physical and mental pressures, often for very dubious reasons. The growth curves of females and males are similar until the onset of the female's growth spurt at about eleven years of age, followed by the males which occurs on average a few years later. Thus there is a period when girls may well be as large and strong as boys of their own age. This is put forward as an argument in favour of mixed sports at this stage.

Perhaps the major factor to be taken into consideration with children in sport is that of bone growth and ossification. Whilst children's bones and muscles are still growing they are mechanically inefficient, so certain actions are relatively harder for them, and overuse injuries can have long term effects. However their flexibility is high and should be maintained into maturity. There are strong arguments for restricting strength work to body resistance exercises only, until ossification is complete *(which occurs earlier in females)* to avoid damage to growth plates. Although light work with weights might be introduced in the mid-teens, the spine should not be loaded until maturity, as torso strength develops late in the growth spurt and the risk of injury must be considered.

Children have a larger surface area to volume ratio than adults, and as a result of the subsequent greater loss of heat over the skin have a faster metabolic rate.

Children's breathing is less efficient than in adults, and they have to breathe a greater volume of air than adults to obtain a given volume of oxygen, and tend to increase breathing rate rather than depth. The resting metabolic rate in the growing child is higher than in the adult, therefore their resting oxygen uptake is closer to their maximum oxygen uptake, than it is in adults. Therefore children have a smaller range or 'metabolic scope' within which to adjust to the greater demands of exercise.

Due to a smaller stroke volume, maximal heart rates can be as high as 220 bpm *(compared with about 200 bpm in young adults)*, with a corresponding reduction in efficiency of action. They also have lower haemoglobin levels than adults, a lower anaerobic capacity, and generate less lactic acid than after puberty.

In children there is no marked difference in $\dot{V}_{O_2}$ max between the sexes, and training has very little effect on it. This changes at puberty under the influence of the adult sex hormones, when as a result of the raised testosterone levels males develop a greater $\dot{V}_{O_2}$ max than females, along with all the other differences seen.

### 5.4.2 AGEING.

Past the physical prime at about 25 years of age, there is a progressive deterioration in physical *(and some mental)* capacities, as a result of the decreasing efficiency of the body's processes. Many of these are reflected in the decrease in fitness measures that occur with age.

Extra collagen fibres are laid down between the muscle fibres of the heart, resulting in a decrease in elasticity and contractility, which reduces the stroke volume and therefore the cardiac output at a given heart rate. Also the maximum heart rate decreases at a rate equivalent to a reduction of about 1-2 beats per year. Both of these changes have the effect of reducing the $\dot{V}_{O_2}$ max and therefore powers of endurance. Extra collagen fibres are also laid down in the lungs decreasing elasticity which reduces lung volumes and therefore gas exchange capacities. Furthermore there is an increase in the cross linkages that form between collagen fibres, so that collagen throughout the body, including tendons and muscle sheaths becomes 'stiffer'.

As a result of progressive bone loss past the age of about 40, increasing amounts of calcium are deposited in the walls of the arteries, which along with the loss of elastic tissue, and the deposition of calcified plaques of cholesterol *(atherosclerosis)*, increases the 'stiffness' of artery walls which increases the peripheral resistance to blood flow and results in a rise in both systolic and diastolic blood pressure. This in turn overloads the heart, and can result in changes in the heart which increase the risk of heart attacks and strokes whilst exercising. This age related 'hardening' of the arteries *(arteriosclerosis)* also reduces the blood supply to organs such as the brain, kidneys, and liver, which interferes with their normal functioning and reduces exercise capacities.

A general reduction in the ability to synthesise proteins results in a loss of lean muscle mass, which is reflected in a 1% loss of strength per year. The loss of muscle mass reduces the basal metabolic rate *(BMR)* or energy demand of the body at rest, so that without a decrease in food intake body fat increases, even if the weight remains steady. These changes decrease exercise performance.

A loss of nerve cells in the brain affects all aspects of body functioning and therefore exercise, and a loss of motor neurones results in the loss of the control of complete motor units.

A general reduction in the efficiency of all homeostatic mechanisms, especially that of temperature control, severely affects all aspects of exercise performance.

Appropriate, regular, graded exercise has a beneficial effect on all those systems that deteriorate with age, and it is important to maintain appropriate activity at all ages.

## 5.5 EXERCISE AND HEALTH

Although some areas are still controversial, there is an accumulation of reliable scientific evidence that aerobic exercise of the correct type promotes what is generally understood as good health, both physical and mental, in young and old alike *(although not necessarily longevity)*.

Although receiving widespread publicity, fatalities whilst exercising aerobically are rare, and invariably the result of some pre-existing condition; if heart disease is present then the risk of death is increased by exercise.

The activity of the immune system may be reduced for a period of up to 24 hours after exhaustive exercise, opening the way for infection. Care should also be taken to avoid over-exertion whilst infected, as this can lead to long term *(chronic)* conditions such as 'glandular fever', post-viral fatigue syndrome, or M.E. *(Myalgic Encephalomyelitis)*, or even heart disease.

There is also the possibility of overuse injuries to the musculo-skeletal system. However, regular carefully graded aerobic exercise in itself carries little or no risk, and brings many benefits. The most widely appreciated of these benefits is that of the decreased risk of cardio-vascular disease.

There are many risk factors associated with cardio-vascular

disease. Several are not related to exercise and are not dealt with here, e.g. inherited disposition, personality 'type' *(also with an inherited element)*, stress, diet, smoking, and some forms of oral contraceptives in females. Others are related to exercise, such as obesity, hypertension *(high blood pressure)*, high blood cholesterol, and diabetes or high blood glucose.

Obesity is inter-related with other risk factors including hypertension and high blood cholesterol, but just being overweight in itself carries an increased risk of heart disease. In association with a properly balanced diet, aerobic exercise can reduce obesity and help maintain an ideal body weight, for example cycling for one hour can use about 600 Calories (2520 kJ), and jogging for one hour can use about 1000 Calories *(4200 kJ)* - the equivalent of about 5 ounces (142 g) of body fat. A reduction in obesity also relieves stress on joints, increases flexibility, and can have psychological benefits.

Hypertension or high blood pressure can be related to stress, smoking, obesity, or be of unknown cause, gradually increasing with age. Hypertensive subjects have a greater risk of cardio-vascular disease and strokes. Aerobic exercise dilates the arterioles supplying the large muscle masses of the body, and by thus decreasing the peripheral resistance, decreases the blood pressure.

High blood cholesterol, or more accurately a high ratio of total blood cholesterol to high density lipoprotein *(HDL)* cholesterol, is associated with an increase in fatty deposits or plaques in the walls of blood vessels and the risk of coronary heart disease. Aerobic exercise decreases this ratio and therefore the risk, but whether aerobic exercise can actually reduce the fatty plaques once they are formed is not clear.

Due to poor diet and inactivity, these cardio-vascular disease risk factors of high blood pressure and high blood cholesterol ratios are increasingly being seen in childhood.

High blood glucose, as found in diabetes, is related to an increase in risk of cardio-vascular disease, which is the major cause of death in diabetics. Aerobic exercise, by maintaining the sensitivity of the tissues to insulin, plays a part in the prevention and/or capacity to cope with Type II *(mature onset non-insulin dependent)* diabetes, which is caused by a decrease in the sensitivity of the tissues to insulin, and not a deficiency of insulin.

Aerobic exercise increases the blood flow through the coronary arteries supplying the cardiac muscle of the heart wall, thus lowering the risk of a blood clot forming in these vessels *(coronary thrombosis)* and blocking the flow of blood and therefore the supply of oxygen and nutrients to the muscle fibres, which in the case of cardiac muscle cannot respire anaerobically. The risk of thrombosis is further reduced as a result of aerobic exercise increasing the rate at which any clots present are destroyed *(fibrinolysis)*, and inhibiting the aggregation of blood platelets which are involved in the formation of blood clots.

The effect of long term aerobic exercise in lowering the heart rate during sub-maximal exercise and at rest, reduces the demands on the heart. It also improves the stability of the electrical activity of the heart, which decreases the risk of life threatening disturbances in their pattern *(as seen in the ECG trace)*.

In summary, vigorous aerobic exercise at least three times a week reduces the risk of a 'heart attack' by at least 50%.

Exercise of all types has a beneficial effect on the musculo-skeletal system.

Muscles, tendons, ligaments and bones all weaken as a result of disuse. During prolonged inactivity up to a third of the calcium salts which are responsible for the hardness of bone can be lost in a process of demineralisation, leading to the weakening of the bone known as osteoporosis. Exercise protects against this bone loss, and can even more than compensate for the loss of bone mass that occurs past the age of about 40, and in post-menopausal osteoporosis in females. Exercise can also increase the tensile strength of tendons and ligaments, and as a result of the release of growth hormone stimulates the synthesis of tissue protein, especially in the muscles.

Exercise also helps in the management and control of various lung conditions, such as chronic bronchitis, asthma and emphysema.

By improving the quality of sleep, and aiding relaxation, exercise has a role in the control of stress, which in turn is risk factor in heart disease.

However, all these beneficial effects of exercise are reversible, and only persist if vigorous exercise is maintained.

## SUMMARY

❖ Fitness and health are not simply defined, it is possible to be fit but unhealthy, and healthy but unfit. Fitness is frequently defined on the outcome of various 'fitness measures' which are neither valid *(ie they do not measure what it is claimed that they are measuring)* nor reliable *(ie they do not provide consistent results)*.

❖ Physical fitness has many different components, with some particularly relevant to health as well as to sport performance.

❖ There are many structural and functional *(physiological)* differences between males and females relevant to differences in sport performance; arising mainly from the higher levels of the male sex hormone testosterone in males.

❖ Children up to the age of puberty should be given special consideration with regard to training, based on a knowledge of their pattern of growth and development, both physical and mental.

❖ In the absence of any reason to the contrary, exercise is considered to be beneficial at all ages, with activity opposing many of the changes associated with ageing.

❖ Regular, carefully graded, aerobic exercise is generally considered beneficial to health; however, over exertion with inadequate recovery can lead to chronic *(long lasting)* ill health, especially in young adults.

# 6 TRAINING PRINCIPLES

---

## OBJECTIVES

To enable the reader to understand the following;

❖ Training principles applicable to most sports

❖ Continuous or steady state training regimes and their use

❖ The use and limitations of training 'formulae' and training zones

❖ Intermittent training regimes and their use; interval training, strength training, circuit training, stage training, skill training, and mobility training.

---

## 6.1 INTRODUCTION

The main principles of training, especially for strength and endurance, are specificity, isolation, progressive overload, recovery, reversibility; along with considerations of intensity, duration and frequency.

Training effects are very specific to the system being particularly stressed, and training is more effective if it is concentrated in such a way that the system being trained is isolated from others, especially when training individual muscles or muscle groups. Also, for the body to increase its capacities, that is it's 'fitness', it is necessary to expose it to progressive overload, interspersed with adequate recovery periods. Once the demand is reduced or removed altogether, the training effects are seen to be reversible as 'fitness' is lost. The deconditioning effects of inactivity include a decrease in $\dot{V}O_2$ max, lactate threshold, blood volume, skeletal muscle mass and contractile strength; and increase in resting heart rate.

Training involves a fine balance between progressive overload and adequate recovery. Overtraining, where there is insufficient rest to allow recovery from training, leads to a decrease in athletic performance; just as would occur with insufficient overload and too much rest. This is why training schedules followed slavishly in the absence of a personal coach may be counterproductive. The main role of any coach is to know the athlete's response to the demands of training.

Adequate recovery is usually ensured by the **periodisation** of training over various periods of time or cycles. For example one method used in training endurance athletes in particular, is for three hard weeks to alternate with an easier one, for three hard months to alternate with an easier one, and for three hard years to alternate with an easier one. However, getting dedicated sports people to reduce their efforts, can be as difficult as motivating non-athletes to increase theirs.

All fitness training regimes can be designed to be **sport** or **game specific**, that is to take account of the special demands of a particular sport or game, for example running, swimming, cycling, rowing, canoeing, skiing etc. Game specific training incorporates the practice of skills whilst carrying out training of the fitness components, for example footballers would perform shuttle runs whilst trying to control a ball.

## 6.2 CONTINUOUS TRAINING REGIMES

Continuous running, swimming, or cycling etc. trains the aerobic capacity and increases endurance. Continuous 'steady state' effort, can be maintained for a considerable time at a certain intensity or percentage of the maximum aerobic power ($\dot{V}_{O_2}max$), in which the oxygen uptake matches the oxygen demand. During maximal work the cardiac output and oxygen uptake can be lower than at a lower work rate, also anaerobic processes are stressed which is not the prime objective. Therefore if the intensity is pitched at the right level, training at a sub-maximal effort is easier, and more effective in training aerobic capacity, than at maximal effort. One simple way to judge a manageable pace for aerobic training is to exercise just beyond that intensity at which conversation stops.

A more accurate measure is obtained by monitoring the heart rate/pulse, as it is a measure of the work being done by the heart. Karvonen recommended continuous effort at an intensity of about 66% of the heart rate reserve, that is the difference between the resting and the maximum heart rate.

*(Maximum heart rate – Resting heart rate)* × 2/3 = X
**Resting heart rate + X = Aerobic endurance training level.**

This represents a training intensity of about 50% of the $\dot{V}_{O_2}max$.

The maximum heart rate can be estimated theoretically by subtracting one's age from 220. However, it must be remembered that endurance training results in a slowing of the resting heart rate *(bradycardia)* and a lowering of the maximum heart rate. So that a twenty year old well conditioned male endurance athlete could find it difficult to raise their heart rate to 200 bpm, whatever their level of exertion.

As far as general endurance fitness is concerned, exercising at this level for thirty minutes three times a week, is recommended as being effective in promoting gains in aerobic endurance.

With the increasing availability of reliable lightweight heart rate monitors, much interest has been generated in using heart rates as indicators of training intensities.

### ❖ *Training and Performance Applications*

*Heart rate is used extensively in sport and exercise as a marker of exercise intensity. This is commonly expressed as a percentage of maximum heart rate (e.g. 70% of max HR) or simply as a rate (160 bpm). Maximum heart rate is variable and dependant upon the individual and does not indicate good or bad abilities. Maximum heart rate reduces with age, leading to the suggestion that the formula 220 minus age predicts maximum heart rate. However, this equation gives results that are inaccurate to the actual maximum heart rate by +/- 10 beats/min. Also endurance training results in a lowering of the maximum heart rate. True maximum heart rate can be ascertained simply through a bout of maximal exercise. Two minutes of exercise (e.g. uphill running is a very effective mode of exercise) performed at 100% of perceived maximum, provides a sufficient stimulus to raise heart rate to a maximal level. Heart rate can be recorded by short term hand measurement of pulse or with heart rate monitors.*

*Working at 60-70% of max heart rate is a good level for sustained low intensity or active recovery training of between 30-100 mins. A higher intensity of 70-80% of max heart rate is an optimal level for endurance training, especially improving skeletal muscle oxidative capacity (see section 1.4). This intensity is referred to as **upper steady state exercise** because it normally coincides with workloads that are just below the lactate threshold. Upper steady state training is typically performed for 30-70 minutes. Work at 80-90% of max heart rate is much more challenging due to the increased lactate levels, that occur when training at or slightly above the lactate threshold. 'Threshold' training is performed for between 20-40 mins and is associated with improving the lactate threshold and thus the percentage of $\dot{V}O_2$ max at which lactate threshold occurs. Training close to the maximum heart rate in an interval manner (see below) is associated with improvements (albeit limited) in the $\dot{V}O_2$ max.*

## 6.3 INTERMITTENT TRAINING REGIMES

These involve alternating periods of effort and recovery, and are based on the principle that the body can in this way be exposed to a greater total work load than it could be during a continuous period of exercise, when various limiting factors come into play. Almost limitless variations are possible with these regimes, as the number, duration and activity of the work and rest intervals can be altered to suit the aim of the training.

**Table 6.1:** *Interval training regimes for different fitness 'compartments'*

| 'Compartment' | ATP/PC Lactic |
|---|---|
| **Training regime** | 4x[4 x 50m] |
| **Work : Relief** | 1:3 |

| 'Compartment' | ATP/PC/Lactic anaerobic |
|---|---|
| **Training regime** | 2 x [4 x 400 m] |
| **Work : Relief** | 1 : 2 |

| 'Compartment' | Lactic anaerobic/aerobic |
|---|---|
| **Training regime** | 2 x [2 x 800 m] |
| **Work : Relief** | 1 : 1 |

| 'Compartment' | Aerobic |
|---|---|
| **Training regime** | 1 x [3 x 1000 m] |
| **Work : Relief** | 1 : 0.5 |

### 6.3.1 INTERVAL TRAINING.

Interval training can improve both aerobic and anaerobic powers, but its use as anaerobic conditioning is more effective after an extended period of aerobic conditioning.

In running, the original Gerschler-Reindell schedules were for a warm-up period that raised the heart rate to 120 bpm; then for a certain distance to be run in a given time to raise the heart rate to 170-180 bpm; followed by a recovery interval of less than 90 seconds in which the heart rate dropped back to 120 bpm. This sequence was repeated according to the aims of the session. If the heart rate did not decrease to the correct level in the allotted time, then this was taken as an indication that the work interval was too demanding. An example for running would be work intervals of 100 m, 200 m, 300 m, and 400 m, but other variations can include work intervals of greater distances.

The intensity of the efforts will determine the length of the recovery interval and the number of intervals possible in one session. A greater number of intense efforts can be achieved if the session is divided up with longer rest intervals between sets of work and rest intervals. Progression can be achieved by increasing the number of intervals and/or sets, and increasing the frequency of sessions per week or month, according to the demands of the event. Again with reference to running it is suggested that the different 'compartments' of fitness could be worked as shown in Table 6.1.

The duration and intensity of the work intervals would vary from person to person. The activity of the recovery interval also enters the equation. Thus it is suggested that an active running recovery interval prevents the complete regeneration of the ATP/PC system between efforts, so that the anaerobic glycolytic system is used more; an active recovery also aids the removal of the lactic acid accumulated during the work interval. A resting recovery interval does allow the complete regeneration of the ATP/PC system between efforts.

Progression is achieved in different ways depending once again on the nature of the event. For example an 800/1500 m runner looking for speed endurance would mainly progress the intensity or speed of the work interval, whereas a 5000/10 000 m runner, requiring more aerobic endurance, would progress the number and reduce the recovery interval.

As the competition phase is approached the number of work intervals is reduced, their intensity increased, and the passive recovery intervals extended until complete recovery is achieved. This change in pattern results in what is sometimes referred to as 'fast repetitions' . Continued changes in this direction lead to race simulations and time trials.

This reduction in repetitions and increasing intensity is also referred to as the 'Pyramid' system of training, and the same principles can be applied to other forms of training such as strength training, and general conditioning.

Most endurance training regimes now tend to avoid extremes such as pure interval running or long slow distance, and incorporate a wide variety of training stimuli, tailored to suit the individual athlete.

### 6.3.2 CIRCUIT TRAINING

A circuit of exercises is performed at several work stations. Circuits can be designed to train various components of fitness, for example specific muscular strength, anaerobic power, mobility, local muscular endurance, aerobic endurance, and general 'fitness'. The number of work stations, number of repetitions, and length of rest interval between stations can all be adjusted to give a total time of up to 40 minutes. Consecutive stations are designed to exercise different muscle groups so as to spread the fatigue. Circuits can be repeated from 3-6 times, with whatever rest interval is required between them.

Stage training is a variation of circuit training in which three sets of repetitions are carried out at each station with short recovery intervals, before moving onto the next.

### 6.3.3 STRENGTH TRAINING

Strength training can be designed for several categories, such as pure strength, speed, explosive power, and strength endurance. Increasing the resistance will increase pure strength, decreasing the resistance will allow the development of speed, plyometrics (Chapter One, Section 1.9.2, p 22) *is particularly effective in* developing explosive power, and increasing the repetitions will increase the strength endurance.

In weight training the degree of resistance is usually quoted as a percentage of the repetitions maximum *(RM)*, that is the maximum load that can be moved correctly with six repetitions, e.g:

12 x 80% RM for strength gain;

20 x 50% RM for strength endurance gain.

Dynamic movements with resistance include elements of concentric, eccentric, and isometric contractions *(Chapter One, Section 1.9.1, p 21)*

Isometric contractions on their own can be economical with time and yet still be effective, for example five isometric contractions, taking 4 seconds to reach maximum contraction and maintaining it for a further 6 seconds, three times a week, can give significant gains in strength.

Isokinetic work, which maintains maximum demand over the complete functional range of joint movements, requires specialised equipment.

### 6.3.4 MOBILITY TRAINING

All mobility *(stretching)* exercises should be carried out after a thorough warm-up, and smooth movements must be employed so that the stretch reflex is not triggered, and injury sustained.

Stretching exercises are of three types.

**Active stretching** is performed by the action of the person's own muscles, that is by contracting an agonist(s) and relaxing the antagonist(s) as in the sit and reach test.

**Passive stretching** is performed with all the muscles relaxed, and the assistance of a partner in gently pushing or pulling the body or limb into the extended position. *(The special case of* **Proprioceptive Neuromuscular Facilitation** *is dealt with in Chapter One, Section 1.8.5, p 20.*

**Ballistic or kinetic** stretching is performed by utilising the momentum of body parts, such as leg and arm swinging.

### 6.3.5 SKILL TRAINING

Although skills can be incorporated into fitness training as described in game specific training, it is better, at least at first and especially for fine skills, that skills be practiced and perfected before any element of fatigue is introduced. Skills are dulled by fatigue, as is seen towards the end of any skillful game involving physical endurance. Skill training is an aspect of Sport Psychology rather than Physiology, and as such is not dealt with here.

## SUMMARY

❖ Training effects are very specific, and regimes should be designed to isolate the system, include progressive overload and recovery periods; training effects are reversible and the gains may soon be lost.

❖ The heart rate deflection point, that is when the heart rate does not continue to increase proportionally to the effort, is not an accurate measure of the 'anaerobic threshold'.

❖ Alternating periods of effort and recovery allow infinite variety, and a greater total work load than continuous methods.

❖ Circuit training, with a circuit of work stations, can be designed to train various components of fitness and muscle groups.

❖ Strength training involves the use of the correct resistance, and number of repetitions, for the specific training need.

❖ Mobility training must be performed carefully after a full warm-up to avoid the risk of injury.

❖ Skills are initially best trained in the absence of fatigue, as skills become 'fixed' they can then be honed in conditions approaching the circumstances of the competition.

# GLOSSARY

Following are brief explanations of some of the important concepts and terms found in this book.

## Abdomen
That part of the body containing the "viscera", ie the kidneys, liver, stomach, and intestines; separated from the thorax by the diaphragm.

## Acetylcholine
A chemical *(neurohormone)* released from pre-synaptic nerve endings, which diffuses across the synapse and stimulates the initiation of an impulse in the post-synaptic membrane. Is rapidly broken down by the enzyme cholinesterase.

## Acid
A chemical which dissociates *("splits up")* in solution to give hydrogen ions *($H^+$)*. Have a pH less than 7. Neutralised by alkalis *(bases)*.

$$CH_3CHOHCOOH \longrightarrow CH_3CHOHCOO^- + H^+$$

Lactic acid $\longrightarrow$ Lactate ion + Hydrogen ion

## Adenosine Triphosphate (ATP)
A compound formed from ADP + P with energy released from Phosphocreatine *(PC)* and/or the breakdown *(oxidation -either aerobic or anaerobic)* of energy rich substrates e.g. glucose. Stored in all cells, especially muscle fibres. When it is broken down by enzyme action back into ADP + P the stored energy is made available for chemical or mechanical work. All the body's energy use is via ATP, which is continually broken down and resynthesised *(average daily turn-over = body weight)*.

## Adipose tissue
Special tissue within which fat is stored. Found mainly under the skin *(sub-cutaneous)* and around the major organs.

## Adolescence
The period in which a second growth spurt occurs and sexual maturity is achieved.

## Adrenal glands
Literally "on top of the kidneys". Composed of two distinct regions, an outer cortex, and an inner medulla. The cortex secretes adrenal cortical hormones, e.g. sex hormones, aldosterone, cortisol; the medulla secretes adrenaline and noradrenaline, and is closely linked to the sympathetic nervous system.

## Adrenaline
A hormone *(chemical transmitter substance)* released from the medulla of the adrenal glands and from sympathetic nerve endings, which prepares the body for "fight or flight" as a result of a "fright".

## Aerobic Exercise
Exercise during which the energy needed is supplied by aerobic respiration *(oxidation)* of energy rich substrates e.g. glucose, using the oxygen that is breathed in *(fats can only be broken down aerobically)*. Such exercise can be continued for long periods.

## Affinity
Attraction to, 'liking' for; e.g. haemoglobin has an affinity for oxygen, with which it forms oxyhaemoglobin.

## Alactacid *(alactate)* Oxygen Debt *(alactic recovery oxygen consumption)*
The oxygen necessary after exercise to replenish the ATP-PC energy stores, and to resaturate the myoglobin and tissue fluids with oxygen.

## Alkali *(or base)*
A chemical which accepts hydrogen ions, thus neutralising acids. Have a pH greater than 7.

## Amino Acids
Organic acids containing nitrogen. Proteins are made up of long chains of amino acids joined by peptide bonds. The body must be supplied with amino acids in the diet. There are 20 different types of amino acids in proteins of living origin. 'Non-essential' amino acids are necessary for body function but can be produced in the body by interconversions of other amino acids; about 11 so called 'essential' amino acids are not produced

in the body *(at least not at a fast enough rate to satisfy demand)* and must be obtained via the diet. Amino acids excess to the body's needs cannot be stored, and are converted into glucose which is used as an energy source, and urea which is excreted in the urine, and incidently in the sweat *(especially during exercise when the kidneys have a reduced blood supply.*

### Amphetamine
A synthetic central nervous system stimulant related to adrenaline.

### Anabolic Steroids
A group of ergogenic aids *(related to the male hormone testosterone)* that have an anabolic *(protein building)* effect, and to a greater or lesser extent an androgenic *(development of male characteristics)* effect on the body.

### Anabolism
That aspect of metabolism involved in the building up *(synthesis)* of complex substances *(e.g. proteins)* from simpler substances *(e.g. amino acids)*. Requires energy in the form of ATP.

### Anaerobic Exercise *(respiration)*
Exercise that demands more oxygen than can be supplied at the time, and which therefore results in the depletion of ATP-PC stores, and the incomplete oxidation of glucose with the accumulation of lactic acid.

### Anaerobic Glycolysis
The initial stages in the oxidative breakdown of glucose in the cytoplasm of cells and muscle fibres, which does not directly involve oxygen, generates a relatively small amount of ATP from each glucose molecule very rapidly, and which in the shortage of oxygen leads to the accumulation of lactic acid.

### Anaerobic Threshold
The 'point' at which, during exercise, the oxygen supply becomes insufficient to maintain aerobic respiration, so that anaerobic respiration becomes predominant, with the accumulation of lactic acid in the blood. Less used than previously, as overlap between aerobic and anaerobic respiration systems at all times of all types of exercise complicate the idea of a simple 'threshold'.

### Analgesic
Pain killer e.g. aspirin.

### Anoxia
Lack of oxygen in tissues.

### Arterioles
Finer branches of arteries, with relatively narrow diameters, and involuntary muscle in their walls, the contraction of which leads to vaso-constriction, and the relaxation of which leads to vasodilation. When constricted *(narrowed)* there is a greater resistance to the flow of the blood and a raised blood pressure, and vice versa. Lead into the capillary beds.

### Arterio-venous oxygen difference (a-$\bar{v}O_2$ diff)
the difference in oxygen content between the blood entering and that leaving the pulmonary capillaries.

### Artery
Blood vessels carrying blood away from the heart, eventually dividing into arterioles.

### Articulate
To connect by means of a joint.

### ATP-PC System *(phosphagen system)*
An anaerobic energy system in which ATP is regenerated from the breakdown of phosphocreatine *(PC)*. Muscles performing at maximal effort obtain ATP from this system.

### Atrophy
Reduction in size and/or mass of cells and tissues, especially relating to muscle fibres.

### Autonomic
Self-controlling; functionally independent of voluntary control.

### Autonomic Nervous System
That part of the nervous system which works involuntarily *(is not under voluntary control)*, controlling all the autonomic processes in the body, e.g. breathing rate, heart rate, peristalsis in the gut, contraction of the bladder, dilation and constriction of the pupil of the eye. Consists of two opposing *(antagonistic)* sub-systems, the sympathetic and parasympathetic nervous systems.

### Basal Metabolic Rate
The rate of the metabolism, as measured by the energy output of an individual, whilst at rest in optimum conditions 12-18 hours after eating *(post-absorptive period)*.

**Biopsy**
The extraction of small pieces of tissues for chemical and/or histological studies, e.g. muscle biopsy to study fibre composition, using a hollow needle.

**Blood Pressure**
The pressure exerted by the blood on the wall of a blood vessel, a function of cardiac output and peripheral resistance *(the resistance to flow of the blood in the blood vessels, mainly the arterioles)*.

**Bradycardia**
Resting heart rate slower than average.

**Buffering capacity**
The capacity to prevent changes in pH.

**Buffers**
Substances which can prevent rapid changes in pH *(acidity and alkalinity)* within the body, e.g. proteins in the plasma, and haemoglobin in the red cells of the blood.

**Calorie**
A unit of heat. A thousand so-called small calories equals one large Calorie *(kilocalorie or kcal)*, which is the type used when speaking of human nutrition. 1 Calorie = 4186 joules (4.186 kJ).

**Carbohydrates**
Organic compound containing only carbon, hydrogen and oxygen in a characteristic ratio, e.g. starch, sucrose *(table sugar)*, and glucose. They are a basic source of energy, circulating as glucose in the blood stream, and being stored as glycogen in virtually all body tissues, but mainly in the liver and muscles. Bread, potatoes, fruits, honey and refined sugars are all excellent sources of carbohydrates. Carbohydrates yield about four Calories per gram when oxidised.

**Cardiac Output**
The amount of blood in dm³ (litres), pumped by the heart per minute, a function of heart rate and stroke volume. Generally the outputs of the right and left ventricles are the same.

**Catabolism**
That aspect of metabolism involved in the breakdown of complex substances into simpler substances. For example the oxidation of glucose into carbon dioxide and water *(with the release of energy)* in aerobic respiration.

**Central nervous system**
The brain and spinal cord.

**Chemoreceptors**
Receptors sensing changes in the chemical composition of body fluids e.g. blood glucose levels.

**Complete Protein
(protein of high biological value)**
Protein that contains all of the essential amino acids, e.g. eggs *(which contain them in the ratio closest to that of human requirements)*, cheese, milk, meat, whole grains, and soya beans.

**Concentric contraction**
Contraction of a muscle reducing its length.

**Connective Tissue**
Tissues that provide support and cohesion for the body, e.g. white collagen fibres which form tendons, the basis of bone, and fibrous cartilage; yellow elastic fibres which form ligaments, and the basis of elastic cartilage; bone and cartilage. Others form sheets or mesenteries which hold organs in place.

**Core Body Temperature**
The central body temperature, as opposed to that of the limbs, the temperature of which is lower due to their greater surface area to volume ratio .

**Coronary**
Relating to the blood vessels that supply the cardiac muscle of the heart wall *(from their 'crown' like arrangement around the heart)*.

**Dehydration**
Excessive loss of water, during exercise mainly as a result of sweating.

**Diastole**
Relaxation, as in relaxation of the ventricles *(ventricular diastole)*.

**Diffusion**
The net movement of gases or dissolved substances, as a result of their kinetic energy, from regions of their higher concentration to regions of their lower concentration, down a concentration gradient, until equilibrium is reached.

**Eccentric contraction**
Contraction of a muscle whilst the length of the muscle increases, e.g. the contraction of the quadriceps in the front of the thigh whilst running downhill.

**Electrolytes**
Substances that dissociate into ions in solution *(ionize)*. See inorganic ions/mineral salts.

**'Empty' Calories**
Calories obtained from foods such as sugar, which are virtually devoid of dietary essentials like amino acids, vitamins and minerals.

**Endocrine glands**
Ductless glands that produce and release *(secrete)* hormones directly into the blood, e.g. pituitary gland, adrenal glands, thyroid gland.

**Energy**
Energy can neither be created nor destroyed. In metabolism, energy in chemical compounds is trapped eventually in ATP, and then either used in synthetic reactions e.g. protein synthesis in growth, or in the sliding filament mechanism in contracting muscle fibres etc.; ultimately all energy is lost as heat.

**Enzymes**
Complex proteins that are capable of speeding up specific chemical reactions without being changed themselves *(ie organic catalysts)*, name ends in "-ase", e.g. sucrase *(catalyses the breakdown of sucrose)*, and maltase *(catalyses the breakdown of maltose)* etc.

**Epithelium**
A tissue lining a body surface, e.g. the lungs.

**EPOC**
Excess post exercise oxygen consumption, the oxygen taken up after the end of a period of exercise. To be preferred to 'oxygen debt', as not all the extra oxygen taken up after a period of exercise excess to normal needs is 'a debt' as such resulting from under supply during the period of exercise.

**Ergogenic Aids**
Substances, other than naturally occurring foods, that when taken orally or by injection will increase the potential for exercise performance, e.g. anabolic steroids.

**Ergometer**
A stationary cycle used for training or for laboratory tests to measure work performed.

**Fast-Twitch *(FT)* Muscle Fibres**
They have a contraction speed 2-3 times faster than slow-twitch *(ST)* fibres, and are capable of producing more power than ST fibres.

**Fat *(lipid)***
Fat acts as an energy store, contains fat soluble vitamins, provides heat insulation under the skin *(sub-cutaneous)*, and support and protection for organs. Fat supplies about nine Calories per gram when oxidised. Fat can only be oxidised aerobically.

**Fatigue**
A subjective experience, not amenable to objective testing, but clearly understood by all sportspersons.

**Fatty acids**
Long chain organic acids which are one of the end products of the digestion of fats *(glycerol being the other)*, which can be oxidised aerobically as a source of energy, or which can be resynthesised back into fats stored in adipose tissue. Some are essential for certain key metabolic processes e.g. the proper functioning of the nervous system, and must be supplied in the diet *(the essential fatty acids)*.

**Fulcrum**
The axis of rotation for a lever

**Functional residual capacity (FRC)**
The volume of air left in the lungs when the respiratory muscles are relaxed.

**Glucose *(blood sugar)***
The simplest carbohydrate in the body *(a monosaccharide or 'single sugar')*. It may be oxidised aerobically to carbon dioxide and water, or anaerobically to lactic acid. It is the sole source of energy for the nervous system. It may be converted into glycogen or fat.

**Glycogen**
The form in which carbohydrate is stored in the body, mainly in the muscles and the liver.

**Glycolysis**
The first stages of cellular respiration occurring with or without the presence of oxygen, in which glucose is converted to two molecules of pyruvic acid.

**Haemoglobin**
The iron-containing pigment in the red blood corpuscles *(erythrocytes)* that combines with oxygen to form oxyhaemoglobin.

### HDL

High density lipoprotein. Lipoproteins are combinations of fat (*lipo*) and proteins. The greater the proportion of protein (*which is more dense than fat*) the higher is the density of the lipoprotein (*HDL*), and vice versa (*LDL*). The ratio of HDL to LDL in the plasma is of more significance to health than straightforward 'cholesterol levels'. The higher the ratio the "better".

### Homeostasis

The maintenance of constant internal conditions (*mainly of the body fluids*) in the face of changing activity and external conditions, to provide optimum conditions for enzyme activity of metabolism. Controlled by negative feed-back loops, in which any change away from the 'goal state' is opposed. The 'ideal state' is never reached, and the metabolism fluctuates or 'hunts' around the optimum within narrow limits, meaning that homeostasis is a dynamic equilibrium, never a static state.

### Hormones

Chemical 'messengers' secreted by ductless endocrine glands directly into the blood, which in small amounts stimulate specific processes of metabolism in 'target' organs or tissues, usually at a distance from their site of production and secretion.

### Hyperglycaemia

Higher blood glucose level than normal

### Hypertension

High blood pressure.

### Hypertrophy

Increase in the size and/or mass of cells and tissues, especially relating to muscle fibres.

### Hyperventilation

An excessive increase in the rate of breathing, which causes a decrease in the amount of carbon dioxide in the blood, resulting in giddiness, cramps, convulsions, lowered blood pressure, and anxiety. Breathing into a paper bag, or closing one nostril

### Hypoglycaemia

Lower blood glucose level than normal.

### Hypothermia

Body temperature below normal.

### Hypoxia

Low oxygen in the inspired air.

and breathing with the mouth shut, help the blood's carbon dioxide content return to normal.

### Incomplete Protein
### (protein of low biological value)

Protein that lacks one or more of the essential amino acids e.g. much vegetable protein.

### Insulin

Hormone secreted by patches of endocrine cells in the pancreas. Opposes any rise in blood glucose by suppressing breakdown of liver glycogen to blood glucose, and stimulating formation of muscle glycogen from blood glucose. Also has a role in protein synthesis. The actions of insulin are opposed by glucagon and adrenaline.

### Interval Training

A system of training in which intervals of hard exercise are alternated with easier recovery intervals.

### Isokinetic Exercise

Contraction of a muscle at constant speed, whilst exerting maximum tension over the full range of movement at all joint angles, rarely achieved without special equipment.

### Isometric Exercise

Contraction of a muscle in which shortening is prevented, e.g. when straining against an immovable resistance.

### Isotonic drink

Being of the same concentration as the blood.

### Isotonic Exercise

Contraction of a muscle during which the force of resistance to the movement remains constant throughout the range of motion.

### Kinaesthetic Feedback

The provision of feedback from proprioceptors (*internal sense organs*) about the position and movement of the body.

### Lactic Acid (*lactate*)

Formed in exercising muscles under anaerobic conditions. It causes the muscular pain associated with intense exercise. It is not a waste product, as it is oxidised as an energy source when oxygen is available. The alternative term 'lactate' is strictly more accurate, as all acids exist in solution in the dissociated form, that is the molecule of lactic acid 'splits up' releasing positively charged hydrogen ions, and the remainder of the molecule, which is negatively charged, is the lactate ion.

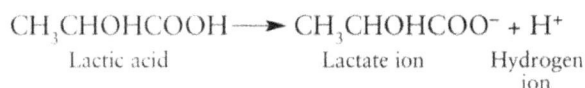

$$CH_3CHOHCOOH \longrightarrow CH_3CHOHCOO^- + H^+$$

Lactic acid   Lactate ion  Hydrogen ion

### Lactic Oxygen Debt
The oxygen necessary after strenuous exercise to remove lactic acid from the blood.

### LDL
Low density lipoproteins, see HDL.

### Ligament
Elastic tissue joining bones to bones.

### Lymph
Plasma, minus plasma proteins, is exuded (pushed) through the capillary walls by the blood pressure and bathes the tissues as 'tissue fluid', which is drained into the lymphatic system, where white cells known as lymphocytes are added by the lymph glands that occur throughout the system. The fluid is now known as lymph. It is returned to the circulatory system in the neck region. The lymphocytes help fight infection, if the lymph glands become infected they become swollen and painful (hence 'glandular fever').

### Maximal Oxygen Consumption
($V_{O_2} max$)
The maximum amount of oxygen that an individual can consume in one minute. The figure may be expressed in $dm^3$ (litres) of oxygen per minute, or more commonly in body weight bearing sports, e.g. running, in centimetres cubed ($cm^3$) of oxygen per kilogram of body weight per minute ($cm^3 \cdot kg^{-1} \cdot min^{-1}$). Remember the dot over the V represents the $min^{-1}$ or 'per minute'.

### Metabolism
All the chemical processes involved in maintaining life.

### Minerals (inorganic ions, mineral salts, electrolytes)
Chemically simple substances that are essential constituents of all cells. Minerals play an important role in water balance (osmoregulation), regulation of blood volume, maintenance of proper acid-base balance, and all body functions eg calcium is essential for muscle contraction, and sodium and potassium are essential for nerve impulse transmission. Mineral salts are lost daily in the sweat and urine and must be replaced through the diet.

### Mitochondria
Microscopic structures (from 0.001 mm - 0.4 mm) in cells and muscle fibres, just visible under the highest magnification of the light microscope. Centres of aerobic respiration using oxygen, regenerating ATP, and producing $CO_2$ and $H_2O$ as end products of the oxidation of glucose.

### Monosaccharide
Literally 'single sugar', the simplest type of sugar molecules e.g. glucose.

### Motor unit
All the muscle fibres supplied (innervated) by a single motor neurone.

### Myogenic contraction
Initiating contraction without nervous stimulation, although nervous stimulation and hormones are involved in co-ordination and determining rate, e.g. cardiac muscle, and involuntary muscle in the wall of the gut.

### Myoglobin
'Muscle haemoglobin', an iron containing muscle pigment, that when oxygenated acts as an oxygen storage compound in Slow Twitch muscle fibres, imparting a red colour, hence red muscle fibres.

### OBLA
Onset of blood lactate accumulation. Although there are normally traces of lactate in the blood, it is generally agreed that a level of about 2 - 4 millimoles per $dm^3$ (litre) represents OBLA, which correlates to the term 'anaerobic threshold'.

### Optimum
The best possible.

### Osmosis
The passage of water from regions of high water potential (pure water or more dilute solutions) to regions of low water potential (more concentrated solutions), across a partially permeable membrane (one that is more permeable to water than to dissolved substances (solutes)), down a water potential gradient until an equilibrium is reached. It is the special case of the diffusion of water. Sea water is more concentrated than blood, therefore if it is swallowed water moves from the blood and tissues by osmosis into the sea water in the gut. Fresh water is less concentrated than blood, therefore when drunk it moves by osmosis from the gut into the blood.

**Oxidative potential**
The ability to use oxygen in aerobic respiration.

**Oxygen Debt**
The amount of oxygen required after muscular activity for the removal of lactic acid and other metabolic products that accumulate when the supply of oxygen is below the needs of the individual, and to replenish various oxygen stores.

**Oxygen Deficit**
The amount of oxygen that the body is undersupplied with during a period of intense exercise, when oxygen consumption does not equal what is necessary to supply all the ATP from aerobic oxidation, during which time energy is partially supplied from anaerobic stores.

**Parasympathetic nervous system**
The part of the autonomic *(involuntary)* nervous system responsible for promoting normal relaxed functioning. Antagonistic to the sympathetic nervous system, e.g. the sympathetic nervous system stimulates an increase in the heart rate, and the parasympathetic nervous system decreases the rate.

**Partial pressure**
In mixtures of gases, e.g. air, each substance exerts a partial pressure proportional to its concentration in the mixture. This pressure arises from continuous random movements that all gas particles exhibit.

**pH**
A measure of acidity or alkalinity, pH 7 is neutral, increasing acidity is expressed as a number less than 7; increasing alkalinity as a number greater than 7. The normal pH of blood plasma is 7.35-7.45.

**Phosphagen system**
The energy system involving ATP and phosphocreatine *(PC)*. Stores of ATP and PC are exploited first in explosive exercise in what is known as the alactic anaerobic system.

**Phosphocreatine *(creatine phosphate)***
Energy rich phosphate containing substance used as an immediate source of energy in the regeneration of ATP. Phosphocreatine itself can only be regenerated when there is an excess of ATP.

**Plyometrics**
Maximum concentric effort made immediately following an eccentric phase. In simpler terms bounding, hopping, and rebound jumping.

**Power**
The rate of doing work; the rate of transfer of energy. It is defined in watts *(W)*. 1 watt = 1 joule per second.

**Proprioceptors**
Internal sensory organs found in muscles, joints and tendons, which detect movements and position of the body.

**Protein**
Large molecules composed of long chains of amino acids *(see also amino acids)*. Essential for growth and repair, but also a source of energy with one gram of protein supplying four Calories when oxidised. Excess protein *(amino acids)* cannot be stored, therefore daily intake required.

**Pulse rate**
The rate of the pressure waves generated in the arteries as a result of the contraction *(systole)* of the left ventricle. In normal, healthy individuals, pulse rate and heart rate are identical.

**Reliability**
A measure of whether a test gives repeatable results.

**Residual volume *(RV)***
The volume of air left in the lungs after a forced maximal expiration.

**Respiratory Exchange Ratio *(RER)***
The ratio of carbon dioxide produced and oxygen used. Indicates the type of fuel being used in the activity, e.g. aerobic oxidation of glucose *(RER = 1)*, fats *(RER = 0.7)*, and protein *(RER = 0.8)*.

**Respiratory Quotient (RQ)**
See Respiratory Exchange Ratio.

**Sarcomere**
The functional unit of a muscle myofibril, consisting of overlapping actin and myosin filaments between two Z discs *(bands)*.

**Slow-Twitch *(ST)* Muscle Fibres**
Contract at a rate 2-3 times slower than fast-twitch *(FT)* fibres, but have greater endurance. Also known as red fibres as a result of the presence of myoglobin.

**Stationary Air**
The air remaining in the lungs during quiet tidal breathing.

**Strength**
The force that a muscle can exert in one maximal effort.

**Stroke volume (SV)**
The volume of blood ejected by each contraction *(systole)* of the ventricle, is calculated by dividing the cardiac output by the heart rate.

**Superficial**
On or near the surface, visible or palpable *(able to feel with hands)*.

**Sympathetic nervous system**
The part of the autonomic *(involuntary)* nervous system responsible for preparing the body for action *(see adrenaline)*. Antagonistic to the parasympathetic nervous system.

**Systemic circulation**
The general circulatory system of the body, as opposed to that of the lungs *(pulmonary circulation)*. Blood passes through the heart twice, as it flows from the systemic to the pulmonary and back to the systemic circulation.

**Systole**
Contraction, as in ventricular systole.

**Tachycardia**
Resting heart rate faster than average.

**Temporal summation**
An increase in responsiveness of a nerve or muscle fibre, resulting from the additive effect of frequently occurring stimuli.

**Tendon**
Tough, non elastic fibrous tissue attaching muscles to bones.

**Testosterone**
The male sex hormone secreted by the testes in the male, and by the adrenal cortex in both males and females. Responsible for the development of male characteristics.

**Thorax**
That part of the body containing the heart and the lungs, separated from the abdomen by the diaphragm.

**Tidal volume**
The volume of air moved during breathing .

**Total lung capacity**
Vital capacity + residual volume, difficult to measure.

**Validity**
A measure of whether a test actually tests what it claims to test, e.g. does the Conconi test give a measure of the anaerobic threshold?

**Vasoconstriction**
A decrease in the diameter of a blood vessel *(usually an arteriole)* by contraction of circular involuntary muscle fibres in the walls, resulting in a reduction of blood flow to the area supplied by the vessel.

**Vasodilation**
An increase in the diameter of a blood vessel *(usually an arteriole)* resulting in an increased blood flow to the area supplied by a vessel.

**Vasomotor**
Relating to the control of vasoconstriction and vasodilation.

**Viscosity**
'Thickness' of a fluid or 'ease of flow', e.g. plasma has a viscosity which allows it to be pumped rapidly around the body.

**Vital Capacity**
The total volume of air that can be expired following full inspiration, in other words the total volume of air that can be moved over the lungs in one 'breath'.

**Vitamins**
Complex organic substances required in the diet *(NB vitamin D also produced by the action of ultra-violet light on the skin)*, essential for normal body functions and maintenance of health. Vitamins contribute to the regulation of metabolic processes, including a role in energy transformations.

**Work**
Application of a force through a distance; it is measured in joules *(J, kJ, MJ)*.

**Work rate**
Work performed per unit time = power.

# INDEX

www.ingramcontent.com/pod-product-compliance
Lightning Source LLC
Chambersburg PA
CBHW051347200326
41521CB00014B/2505